LONDON ILLUST

The Definitive Guide

With around 200 colour photographs, illustrations and maps, the guide
traces the story of London over many centuries and provides an
A to Z companion for Londoners and Visitors alike.

Professor S K Al Naib

St Paul's Cathedral

City of London

Welcome to London - Your Illustrated Guide

The Guide Book

This comprehensive guide is designed to help Londoners and visitors enjoy the old and new in London. It makes a point of covering not only the established historic attractions but also the developments which are likely to be centres of interest in the 21st century. Two approaches are included: a geographical account and a theme listing. In the first major part of the book there are many chapters, beginning with a brief history of London from the Roman times through the various ages up to the present day. Ten sections are covered in detail to depict the attractions in various parts of London.

Attractions & Heritage Locations

The starting point is the City of Westminster which is the centre of British government and the main home of the monarchy; in the succeeding sections the surrounding areas are dealt with and finally attention is focused on the River Thames. There are aerial photographs and clear maps with travel information. Each page concentrates on one or two features and the surrounding buildings. On each item there is a short history with architectural details. Where possible the illustrations provide a comparison of past and present and in some cases a glimpse of the future.

Quick Theme Reference

In the last part of the book there is a quick reference to those sights and events which may be of interest to the reader. The information on ceremonies and other events is believed to be accurate at the time of publication but the reader is advised to confirm details. For this purpose, contact telephone numbers are included where possible. Please remember to check admission times and charges before setting out. Visitorcall (information) telephone numbers are also given on p133.

Great History - Enjoy London's 2000 years concise history (p6-16).

Choice of Heritage Location - Wherever you want to go in the big city or village London, there is certain to be a welcoming attraction (p17-101).

Historic London Attractions include Tower of London, Buckingham Palace, Westminster Abbey, Houses of Parliament, etc.(p18).

River Thames Attractions - Do not miss the wide choice of river excursions, including a panoramic tour of Central London (p103).

London's Subterranean World Descend down the London Water Ring Main shaft or track an underground river (p110).

Royal Palaces and Castles London has the very best collection of Royal Heritage the country has to offer (p117).

The Great Art Treasures Take advantage of the London galleries and museums (p118).

Colourful Ceremonial Events - Enjoy with the children, the Changing of the Guard or the Lord Mayor's Show (p121).

Walking Tours - Choose the freedom of the road with city heritage walks (p122).

Antique and Street Markets - Complete your enjoyment of the capital with a souvenir from one of the old street markets (p124).

Shopping Around - There is lots to buy for the family in the world famous Oxford Street, Harrods and Knightsbridge (p125).

Special Sport Interest - Indulge your passion for sport with a choice of world events at well known venues (p126-27).

Nature Lovers Breaks - London abounds in parks, gardens and boasts the largest Zoo and collection of city farms (p129).

Pubs and Ale Trails - Celebrate or simply indulge in trails around the city historic inns or waterside pubs (p130,69).

London Music and Theatre Breaks Choose from hundreds of conveniently located London cinemas, concert halls and theatres (p131).

Easy to Read Maps - Over 20 maps indispensable to the wandering visitor or Londoner.

Aerial Photographs - Admire more than 50 superb bird's eye views of famous landmarks in all parts of London.

I wish you a happy and enjoyable visit to this ancient city that was once the heart of the British Empire stretching over five continents, and is now one of the world's foremost modern financial centres and a custodian of much that relates to Great Britain.

ISBN 1 874536 01 5
First Printing March 1994

Books by the Author

"London Illustrated" Historical, Current and Future. ISBN 1874536 01 5

"Discover London Docklands" A to Z Illustrated Guide ISBN 1 874536 00 7

"Hydraulic Structures" Theory, Analysis and Design ISBN 0 901987 83 2

"London Docklands" Past, Present and Future ISBN 0 901987 83 6

"European Docklands" Past, Present and Future ISBN 0 901987 82 4

"Dockland" Historical Survey ISBN 0 901987 80 8

"Applied Hydraulics" Theory and Worked Examples ISBN 1 874536 02 3

"London History and Heritage" 2000th Year Anniversary Guide (Forthcoming)

The author is Professor of Civil Engineering and Head of Department at the University of East London, England. (Tel: 081 590 7722)

Printed by The KPC Group, Kent.

Historic Tower of London

Preface - The Making of Modern London

Big Ben

A remarkable aspect of London is that its history and development have controlled the destiny of the United Kingdom and many countries all over the world. This book presents to the visitor a portrayal of its rich history and sets the scene for many rewarding tours. It covers the most important of the historic houses, Royal Palaces, notable buildings, monuments and riverside attractions. Here are some particularly fine and interesting examples of architecture over the last five centuries which form an intricate part of the story of London.

Over the past quarter of a century the face of London has changed dramatically and this change is particularly apparent along the course of London's greatest highway, the River Thames. New communities have emerged on the riverside as the once thriving port with its industrial activities has given way to stylish residential, commercial and leisure related developments. New commercial centres such as Canary Wharf, London Bridge City and other riverside developments have created areas of new employment. The vitality and vigour of the new London proves that the city is on the move. Riverside renaissance, beauty and charm will amaze you.

The city that the Victorians viewed at the end of the 19th century now presents a different but just as inspiring panorama. Among many new shapes on the skyline are the Telecom Tower, representing a means of communication never dreamed of before, skyscrapers and multi-storey buildings in the commercial district of the City and in Docklands beyond, and a dozen refurbished railway stations, from which millions of London's commuters emerge daily on their way to work in the capital. The nature of work in the metropolitan city has been transformed. Indeed many inventions have changed the lives of Londoners. Gas followed steam, then electricity, the cinema, TV, X-rays, electronic and information technology. Today, engineering science and new concepts in satellite communications offer great challenges. London is being changed in many tangible ways and its influence will continue to dominate all parts of Great Britain and Europe.

This book will give the reader an insight into where London might be in the 21st century - not only changing its outward appearance through development projects, but providing the environment to make it more competitive as a financial centre of the world, more vibrant as a national communication focus, more convenient to travel around and more enjoyable for those who live and work in this historic capital city.

London is one of the most popular tourist destinations in the world. It now plays host to over ten million visitors a year. Approximately half come from Western Europe and a third from the United States and Canada. A major strength of London as a tourist centre is that it offers diverse attractions as well as acting as a gateway to other areas of Great Britain and most of Europe. Many visitors consider London as an essential part of their holiday package whatever their plan of travel.

What is special about London is that history has left a considerable amount of well preserved heritage visible throughout the city. You can see clearly the contributions and marks of each passing century since Roman times, which means you can sense the spirit of each period in its history. This is a very important and unique quality of London when compared with other major cities of the world today. The intention of this book is to recreate the old scenes as an expression of the history of the city over many centuries. Another characteristic of great value is that London consists of distinct boroughs and communities with their specific identity and heritage. It is essential that a city district functions locally as well as it functions as part of the City as a whole.

Many sites in Central London have been occupied by a succession of buildings over the past centuries. In 1666 the Great Fire of London destroyed many thousands of houses which were replaced by larger and more functional buildings. During the Blitz of 1940, large parts of the City were destroyed and have since been rebuilt as tall modern office complexes set amongst the historic landmarks. Whilst most British towns and cities developed primarily as manufacturing centres during the Industrial Revolution, London owes its origins to the city trade and business. Offices have always been in demand and they are continuously being built. Some of the office buildings recently completed attempt to preserve and enhance the character of the city.

Vision for the Capital focuses on its form and nature, assessing what lessons can be learnt for the future development and success from changes that have taken place over the past decades. Key issues include: how the shape and role of the city are changing; action needed for the public realm and infrastructure; ways to increase the take up of surplus office space and any special planning principles needed for particular sites. The process of change is set to continue into the 1990s and beyond into the 21st century to maintain London as one of the leading cities in the world.

Contents

Yeoman Warders, Tower of London

Central London and Greater London Boroughs

Central London

Central London is the historic heart of the City. It is the centre of government, banking, finance, commerce, insurance, shipping, entertainment and the arts of the nation with worldwide influence. Developed over the past two thousand years, the streets, squares, churches, palaces and buildings are often many hundreds of years old. There are three districts of Central London: the City of London (often called The City), Westminster and the West End.

Westminster

Westminster is the seat of Government in Great Britain. The Houses of Parliament and the various ministries, along with the Queen's main residence of Buckingham Palace are all situated here and they look after the country's affairs. During the 11th century, the Normans developed Westminster as the place of Government and in the reign of Henry III (1216-72) Westminster was established as the home of the Royal Family, the seat of the Royal Court and the meeting place of Parliament. The Strand, in the West End, connects Westminster to the City of London.

West End

The West End extends from Westminster Palace north to Regents Park, and from Holborn west to Hyde Park. It includes St James's Park and Green Park, formerly church lands acquired by Henry VIII through the dissolution of the monasteries. Today, the West End is world famous for its shopping and entertainment.

The City

The City is a collection of businesses each with its own particular interest. Stockbrokers and bankers are based around the Stock Exchange whereas the insurance industry is based around Lloyds. Bankers meet in Threadneedle Street and Lombard Street. The grouping of trades is a fascinating feature of the old City. In the Middle Ages, practically all the trades had their own locality and the names of existing streets tell their own history - Silk Street, Milk Street, Ironmonger Lane, Wood Lane etc. Each street had its own particular trade, Livery Hall and Church. Many of the Livery Halls are still in existence and all are worth a visit. The City is the oldest and yet most modern part of the Capital. Narrow medieval lanes contrast with the wide straight roads of the 19th and 20th centuries. Since World War Two, giant office blocks and skyscrapers have been constructed. Although the resident population in the Square Mile of the City is about 5,500, approximately a third of a million commuters and traders pass through daily. The 'Big Bang' of the mid 1980s has changed some of the businesses. The commerce of the City has expanded eastward.

London Docklands

Docklands in East London presents an exciting mix of old London and 21st century developments. The wet docks that once offered merchants' ships the greatest port in the world now form a backdrop to some of the most exhilarating new residential and commercial buildings in Europe. The world's largest project, the Canary Wharf, with its technological wonders, stands side by side with the last surviving 19th century warehouses and Georgian buildings.

Greater London Boroughs

In 1965, following the recommendations of a Royal Commission, Greater London was redefined, the boundaries drawn up for thirty two Boroughs and the Greater London Council was created, replacing the old London County Council. The GLC doubled the geographical size of the old LCC. A major part of Middlesex disappeared administratively, also parts of Essex, Kent, Hertfordshire, and Surrey. The GLC shared its many responsibilities with the London Borough Councils and functioned for twenty years but was dissolved in 1986. The London Boroughs now function independently as local authorities with their own thriving communities. Londoners have learned to use sections of the M25 Oribital Motorway as a means of getting to or from radial routes connecting all parts of Greater London.

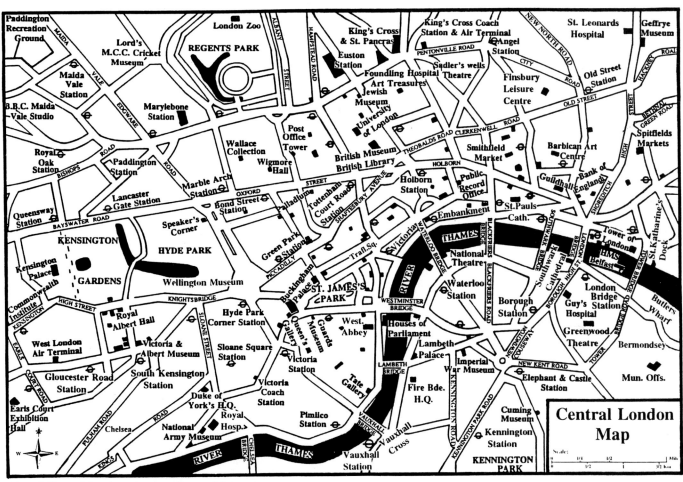

Map of Central London showing places of interest

LONDON HISTORY

The River Thames on Lord Mayor's Day, circa 1747

City of London and surrounding villages, circa 1799

Prehistoric and Roman London 54BC - 410AD

The Thames and the City of London in Roman times

Caesar's Visit 54BC

The City of London is one of the world's great financial centres and until thirty years ago had one of the world's greatest ports. The modern office blocks rest on the remains of a Roman city and port of nearly two thousand years ago. In July 54BC, Caesar reached the Kent shores with a fleet of 800 boats transporting from France an army of five legions, about 17,500 infantry and 2,000 cavalry. He crossed the River Thames probably somewhere near Brentford, opposite Kew Gardens, and attacked the British under their leader Cassivellaunus on their retreat to Verulamium at St. Albans. The tribes surrendered and peace was gained. After a short stay of two months, Caesar and his army sailed back to France.

Roman Conquest 43AD

In 43AD The Emperor Claudius invaded Britain with 50,000 men and incorporated the whole of the south east into the Roman Empire. London (Londinium) at that time was much frequented by merchants and trading vessels. During the next three centuries, London was the largest city in the country and engaged in seaborne commerce. Being the furthest point in the interior to which ships from Europe could penetrate, it became the centre of a Roman system of roads which radiated to other cities.

Since that time the function of the capital city has remained the same. The boundary of the old Roman wall still marks the Square Mile of the City of London and its business community. Built on two hills, Cornhill and that of St. Paul's on the north bank of the Thames, the city was bounded to the west by the River Fleet and to the east by the Wallbrook, both of which are now running underground. A Roman timber bridge connected the city to the south bank, almost on the site of the existing London Bridge. Southwark grew up at the southern end of the bridge, and the Old Kent Road is the Roman road connecting London Bridge to the Kent coast. These and other centres of population served a prosperous countryside dotted with flourishing villas and estates.

Roman Basilica and Wall

By the end of the 1st century London had the first Forum and Basilica with a temple dedicated to Mithras, built nearby. The Basilica was the Civic Centre and the Law Court. The palace was the residence of the Roman Governor of Britain. His troops were stationed at a fort to the north west of the City. Recently remains of the basilica wall were discovered incorporated as a part of the 15th century Leadenhall Market. By 200AD a wall nearly two miles long and 6 metres high had been built around London which determined the shape of London for more than 1200 years and some sections can still be seen in the City today. It was built of ragstone facing with courses of flat bricks and a rubble infill. The stone came from the quarries of Kent. The remains of three Roman boats have also been found in London. The first one in 1910 when the foundations of the County Hall, Lambeth were excavated; the other two barges, discovered at Blackfriars in 1958 and 1962 had flat bottoms suited to carrying heavy loads in shallow waters - one indeed still had on board a load of Kentish ragstone, quarried in the Maidstone area.

Roman Amphitheatre

The discovery of a Roman amphitheatre in 1992 beneath the site of the eastern extension to the City of London Guildhall opens a new significant chapter in the history of London. From what has been found the amphitheatre was similar in design and scale to the structure built by the Romans in Chester. Archaeologists have calculated that the remains could extend under the Guildhall, the modern west wing being built in the 1960s and the St. Lawrence Jewry Church on the south side of Guildhall Park (see page 49).

London Commerce and Trade

The conquest of Britain (Britannia) by the Romans had both political and commercial motives. They encouraged shipping to operate all year round from London. One of the important exports during this period was corn. Another prize commodity which Britain offered was the supply of metal ores from which lead and silver were extracted. Iron and Cornish tin, and gold from Wales were also exported. Other valuable exports included timber, cattle, sheep, hides and wool. On the import side, the presence of a garrison of 40,000, together with a host of Roman officials, created demand for all sorts of imported cargo such as wine, pottery, glass and luxury goods from the Continent. For some three centuries after the Roman conquest, the development continued until the beginning of the 5th century when the Roman empire was weakened by attacks on its borders.

Anglo Saxon and Norman London 410 - 1198

East Saxon 410AD

In 410AD, the Emperor Honorius decided that Britain should be responsible for its own defence. Subsequently London was invaded by seafarers and tribes from the continent who conquered South East England and by the year 600 the Saxon settlement was established. In 601 St. Augustine arrived in Kent on his mission to convert the Saxons to Christianity. The process was largely complete by about 700. Late Saxon England also saw the development of trade within a money economy and the emergence of towns such as Southwark.

London survived behind its Roman fortifications and walls, the traces of which have been preserved along today's London Wall. During this period the city had imports of timber resin and wine in exchange for corn and wool, the island's two principal exports. The city became the capital of the East Saxons and had its own bishop in 604. The Anglo-Saxon chronicler, the Venerable Bede, stated the same year, that "London was the market town of many nations who came to it by sea and land". Trade was established with Marseilles, Rouen, and Troyes. A band of German merchants began to frequent Billingsgate.

The Vikings and King Alfred

In the 9th century the Vikings came to Britain from Denmark, first as pirates and then as settlers. It is recorded that in 851 three hundred and fifty ships sailed into the Thames and invaded Canterbury and London. The Vikings established themselves in London and extended trade with the Baltic countries. Between 850-865 the Danes began to settle in the north east part of Britain. In 883 King Alfred the Great captured London from the Danes and fortified it. He then turned his attention to the development of the shipping business. King Ethelred built the dock on the Thames Ethelredshithe (hithe meaning wharf or landing place) which was subsequently changed to Queenhithe when the property was granted to Matilda, Queen of Henry I in the 12th century.

Norman Conquest 1066

King Harold succeeded to the throne following the death of King Edward the Confessor and this precipitated the Norman invasion. Harold was slain at the ensuing Battle of Hastings, and William Duke of Normandy was crowned in Westminster Abbey in 1066. The population of London was then about 10,000. The citizens were granted a charter, which is still in existence, confirming their rights and privileges. After the Conquest, the natives of Normandy in northern France settled in London and encouraged merchants from Flanders, Spain,

Italy and other European countries to do the same. London became the European centre of trade and the warehouses contained a huge variety of merchandise. We know that merchants, mostly foreign, imported wine and fish from Rouen, and pepper and spices from Antwerp in Belgium - the spices came to Antwerp by camel caravans from the East. Here was the Vintry where the vintners of Rouen congregated. Eastcheap had its Ghent and Ponthieu goldsmiths, at Dowgate were clothiers from Flanders and Cologne. The English merchants and ship owners also ventured overseas and imported silk, gold, garments and dyes, wine, oil, ivory, brass, copper and glass.

The Norman merchants and their descendants, who intermarried with the English, became an important group in the life of London. One of them, Thomas Becket, (1118-1170), served as the Archbishop of Canterbury. Trade in the city grew and in order to control the growth and ensure the collection of the royal tolls, as well as to protect the city, William the Conqueror (1027-87) began to build the stone Tower of London on a little hill commanding the city and the River Thames with its quays and wharves in 1078. The Norman buildings were large and had great strength and scope. They built castles to secure London and the surrounding country against any possible rebellion. Anglo-Saxon galleries are on display at the British and London Museums.

Domesday Book

William the Conqueror wished to know the value of each of the Country Manors held by his barons and sent Royal officers into each county to gather information. Their surveys were recorded in a document called The Domesday Book. You may see it today in the Public Records Office in London, just as it was written in 1086, over 900 years ago.

14th Century Pewter badge flanked by Edward the Confessor and Thomas Becket

Roman Remains and Surviving Early Churches

Prehistoric Sites

Caesar's Camp, Wimbledon Common
Access path through grounds of the Royal Wimbledon Golf Course.

Loughton Camp and Ambresbury Banks Two miles apart in Epping Forest.

Wheathampstead Oppidum, Herts.
Half a mile south east of the town centre.

St. Ann's Hill, Surrey
Hillfort near Chertsey.

St. George's Hill, Surrey
Hillfort near Weybridge

Swanscombe, Kent
Geological formation in a disused chalk pit (now nature reserve) which has produced the earliest human remains in Britain.

Roman Wall and Temple

Corner of City Wall, GPO Yard
By prior permission of the Postmaster Controller.

Foundations of Roman Fort Turrets,
Noble Street

City Wall
In gardens east of Museum of London.

City Wall and Roman Fort Wall
South of St. Giles Cripplegate churchyard and at St. Alphage churchyard.

City Wall with Roman Wall
Below courtyard of 8/10 Cooper's Row.

City Wall and Base of Roman Turret
In gardens south of Tower Hill tube station.

Roman City and Riverside Walls
At Tower of London.

Temple of Mithras
In forecourt of Legal & General Assurance Society, Queen Victoria Street, EC4. Foundations of the Temple of the Roman Sun God, were discovered in 1954 and since being moved to their present position, can be seen in the forecourt at any time. Other finds may be seen in the Museum of London, London Wall, EC2.

Central London

All Hallows by the Tower, Byward Street Surviving Saxon arch.

London Wall Early medieval wall at St Alphage Gardens, part of a signposted London Wall Walk between the Museum of London and the Tower of London.

St Batholomew the Great, Smithfield
Priory Church, mainly 12th Century.

St Bride's, Fleet Street In crypt, foundations of Saxon and Medieval church.

St John's, Clerkenwell
12th Century crypt.

St Mary-le-Bow, Cheapside
Late 11th Century crypt.

Temple Church, Fleet Street
Late 12th Century nave and porch.

Tower of London Late 11th Century White Tower and 12th Century wall.

Westminster Abbey
Late 11th Century Chapel and crypt.

Westminster Hall
Late 11th Century with later roof. This is the only part of the ancient buildings standing today, after nine centuries. It is used on special occasions for important meetings.

Outer London

Barking Abbey, Essex
Ancient Abbey church walls and Curfew Tower.

St Mary Magdalen, East Ham E6
12th Century with 16th Century tower.

St Peter and St Paul, Harlington, Middlesex Norman nave and doorway.

St Mary, Harmondsworth, Middlesex
Norman South aisle and doorway.

Croydon Palace, Surrey
Norman undercroft and Medieval Hall.

Norman abbey churches and castles are at: Waltham Abbey, Essex; St Albans Cathedral, Herts; Rochester Castle and Cathedral, Kent; Berkhamstead Castle, Herts; Windsor Castle, Berks; and the unique Saxon wooden church at Greensted, Ongar, Essex.

Verulamium, St Albans
Roman site with a theatre and museum. The theatre was constructed during the middle of the second century. BR St Albans station is nearby.

Roman Bath, Welwyn
The remains of a Roman bathing suite are on show in a specially constructed vault underneath the A1(M). The suite dates from the 3rd century AD and the layout of the hotroom and baths can be clearly seen. Located at Dicket Mead, Welwyn, off the Welwyn Bypass (0707 271362).

Remains of the Roman Wall near Tower Hill

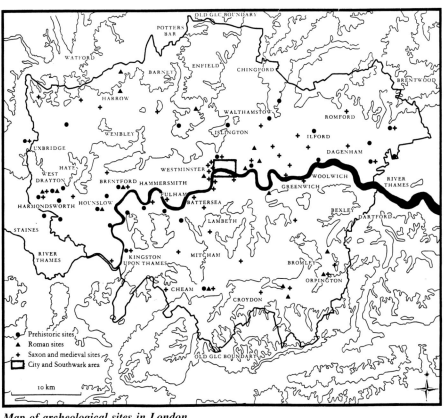

Map of archeological sites in London

Medieval London 1199-1485

Medieval Era

The medieval period ranged over three centuries from the accession of King John in 1199 until the death of Richard III in 1485. During these years there were revolts, uprisings and battles against the Scots and the Welsh and finally the long War of the Roses, at the end of which England emerged as one of the most powerful nations in Europe. The Monarchy borrowed money from the city merchants to fund their wars and to extend British influence overseas. In return the merchants and the city mayors enhanced their status and privileges. Henry III initially refused to accept the election of Thomas Fitz Thomas as Mayor where upon London supported the rebel Simon de Montfort. Having been defeated by the rebels, the King was forced to accept Fitz Thomas as the Mayor.

The Black Death

In 1348 a plague, called the Black Death, started in Asia and rapidly spread across Europe. In England almost half of the population perished. This caused a shortage of labour and the Lords of the Manors could not get enough farm workers.

The Peasants' Revolt

As labour became scarce, the labourers demanded higher wages. Parliament intervened and passed the Act of the Statute of Labourers to fix wages. Many labourers refused to work under these conditions and the Peasants' Revolt broke out in 1381. Under the leadership of Wat Tyler, the rebels marched on London, burned properties and murdered people. Accompanied by the Lord Mayor of London, the young King, Richard II, met the rebels at Smithfield. While the King was speaking to the crowd and promising to improve their conditions, Wat Tyler tried to strike the Lord Mayor but a guard drew a sword and slew him.

First Lord Mayor of London

The first mayor of London Henry Fitz Eylwin was elected in 1198 and he remained in office until his death - 20 years later. By the Magna Carta of 1215 King John granted the City of London a charter and the right to choose a mayor annually. The title 'Lord Mayor' was established in 1414. He had full autonomy within the City walls and yielded precedence only to the monarch. He ranked as an Earl and wore an Earl's scarlet and ermine but was not honoured as such. The City was divided into wards each headed by its alderman. By the late 14th century a common Council composed of ward representatives was established. Merchants and craftsmen formed guilds to control standards of quality, prices and wages. These have survived under the name of Livery Companies and today total about 100.

London's Old Stone Bridge

Upstream from the Tower, where there was a ford across the river to Southwark, the traders decided to build a stone bridge in place of the wooden one that had been burnt down. Built between 1176 and 1209 by Peter de Colechurch (a chaplain of the now demolished Church of St. Mary Colechurch), the stone bridge consisted of nineteen arches resting on stout piers streamlined by boat shaped timber pilings called 'starlings'. Later, houses and shops were built on the bridge along with a chapel dedicated to Thomas Becket where travellers, for safe passage, paid a donation towards maintenance of the bridge. A fortified gatehouse guarded the southern end near Southwark Cathedral, where heads of 'traitors' executed on Tower Hill were exhibited. There was a drawbridge with a toll of sixpence to be raised for ships to reach the upriver wharves!

Londoners and Trade

Much of London's influence and wealth stemmed from its bustling trade which by 1400 reached a high peak. In the warehouses along the Thames corn, cloth and wool were stocked for export, while imports included wines, spirits, pottery and many rare goods from other countries. Medieval Londoners were not only dependent on the Thames for shipping and trade, but also for all their water for drinking and washing. Narrow lanes and streets ran down from Thames Street to the river. Merchants, brewers, tanners and fishmongers lived along these lanes. Hundreds of timber framed buildings were crowded in the lanes and the waterfront.

The Wilton Diptych depicts King Richard II

Old London Bridge by Hollar

Tudor and Elizabethan London 1485-1603

King Henry VIII

Tudor and Elizabethan Dynasty

Towards the end of the 15th century London was still bounded by the old Roman walls, within which most of its 80,000 inhabitants lived. Although it was governed by the Lord Mayor and his Aldermen, the Livery Companies exerted considerable influence. The homes of merchants' gabled houses, the nobility, the monasteries and numerous churches including St. Paul's, crowded the narrow streets. During 1476-77, the Lord Mayor, Sir Ralph Joceline, made the various Guild Companies repair the city walls. However, outside the city, the ditch was neglected. Originally, the ditch, which ran from the Tower of London to Fleet River, was known as Hondesdich which probably derived from the fact that for centuries citizens had thrown their rubbish there and it was the haunt of scavenging dogs. Later the ditch was cleaned out and made into a sewer.

While Henry VII followed his predecessors' custom by staying at the Tower before his coronation, he was also responsible for turning it into a prison for traitors, from which there was usually no escape. The first prisoner was Edward, the Earl of Warwick, who was executed in 1499 at Tower Hill. A scaffold and timber gallows was erected and the nearby Church of All Hallows Barking was where the headless bodies were buried. London's other place of execution was Tyburn, to which criminals were taken from Newgate Jail along Tyburn Way (today Oxford Street). East Smithfield, Wapping and the Isle of Dogs, east of the city, were the haunts of river pirates and places of execution for those that were caught!.

The Tudor period was a turning point in British history. Great names alone reflect just how important. Henry VIII, Elizabeth I and Mary Stewart were among the sovereigns of England and Scotland. Wolsey, William Cecil and Leicester were among the politicians. Shakespeare, Marlowe, Hilliard and Byrd were among the creative artists. John Stow published his important book 'A Survey of London' in 1598.

The Reformation

The 1536 Reformation Legislation Act removed the last vestiges of Papal power in England. The resplendent courts of Henry VIII and Elizabeth, the resolution of Sir Thomas More, the introduction of Church Reformation, the English Bible and the Prayer Book, The Monarch as head of the Anglican Church were among the matters of note. The period saw the development and legacy of Tudor architecture and many of the beautifully designed houses such as Hampton Court Palace remain today. The population of London continued to increase and this stimulated economic growth and increased trade, although certain areas of the city were plagued by famine and disease.

Birth of English Theatre

Many of the famous English theatrical actors and playwrights date from the Elizabethan period, and the most famous of all was William Shakespeare (1569-1616). His plays were presented at the Globe Theatre which was built in 1599 at Bankside, south of the River Thames. This is because the City of

London Fathers disapproved of playhouses. The first English theatre was opened in 1577 by James Burbage in Shoreditch.

Spanish Armada 1588

On 15th June 1588, messengers reached Greenwich Palace with the news that the Spanish Armada was approaching the Scilly Isles. Londoners established a boom across the River Thames at Tilbury and mounted it with a battery of guns to face any ships sailing up the estuary. There were nine batteries placed covering the whole of the river up to Blackwall where there was a second boom. Elizabeth I left St. James's Palace and rode to Tilbury on 8th August to review the troops. In the city more than 10,000 people volunteered to help defend London. The defeat of the Armada was under the leadership of Lord Howard of Effingham, Lord High Admiral, with Sir Francis Drake in command of the fleet.

Legal Quays 1558

From the reign of Queen Elizabeth I almost all cargo had to be landed at "Legal Quays" for the payment of duty. The Legal Quays stretched along the north bank of the Thames from London Bridge to the Tower where customs officials were on duty to carry out the necessary formalities. As trade grew the Legal Quays became unable to cope with the increase in volume and from 1663 "Sufferance Wharves" were established on both sides of the river eastwards from the Tower and along the South Bank from London Bridge.

Map of Elizabethan London by Braun and Hobenburg, circa 1572

17th Century London and the Great Fire of 1666

The Stuart Period 1603-1688

The Stuarts, who came from Scotland, were one of the least fortunate dynasties. Charles I was put on public trial for treason and beheaded. James II, fearing a similar fate, fled the country and abandoned his kingdom. James I and Charles II, died peacefully in their beds. During this period there were two decades of civil war, revolution and republican experiment. 1605 saw the Gunpowder plot and the arrest of **Guy Fawkes** who placed a ton and a half of gunpowder in a vault beneath the House of Lords with the objective of terminating the Scottish Royalty in England. He was caught in the act and tried. The public execution of Guy Fawkes and his fellow conspirators took place outside the Parliament House. Oliver Cromwell (1599-1658) assumed the title of Lord Protector and from 1649 to 1660 England was a republic. However, the Monarchy was restored with limited power, and political parties were introduced. In 1620 the Pilgrim Fathers departed on their religious migration to New England in North America.

The Great Plague of 1665

The population of the City had reached about 130,000 at the end of the 16th century and the spread of building outside the city walls was to lead ultimately to the growth of Greater London. During 1665 the Great Plague caused one in three Londoners to die in the epidemic. It was an infectious bubonic plague, like the Black Death, which killed nearly half the population of England in the 14th century. The Plague was brought to this country through vermin in ships returning from overseas. It spread rapidly in the crowded and poor sanitary conditions of the City of London. Doctors did not have the medicine of today to deal with such diseases, which resulted in a large loss of population in the City. Many of the bodies were buried at Blackheath in South East London. One year later the Great Fire destroyed 60 per cent of the City.

The Great Fire of 1666

On Saturday 1st September 1666, a baker working from a timber dwelling in Pudding Lane in the City, went to bed without damping his fire properly. By the early hours of Sunday the fire had rekindled and burnt the staircase with the flames spreading to the next house. A strong wind carried the fire to the warehouses along Thames Street, where huge quantities of spirit, corn, tallow and coal were stored. The waterfront quickly became a screen of smoke and flames. The narrow lanes leading from the City to the river were congested with escaping Londoners and those armed with leather water buckets fighting the fire. Charles II, advised by Samuel Pepys, ordered the Lord Mayor, Sir Thomas Blundworth, to demolish houses and thus create a fire gap but the houses could not be demolished quickly enough. Charles then ordered that whole streets be blown up by seamen using gunpowder. Late on Wednesday night the flames were finally checked. The city was devastated - over 13,000 houses, 80 churches, 44 halls of the Livery Companies, the Guildhall, the Royal Exchange, and St. Paul's Church were all destroyed. Amazingly, only eight people died during the fire. Trade and businesses were at a standstill as their premises had all been razed to the ground.

Plans for Rebuilding

For rebuilding the City, Charles II appointed a group of six Commissioners, including Christopher Wren, who prepared new building regulations for proportioning houses and paved streets. During reconstruction many streets were widened and houses and government buildings were built in brick. The foundation stone for St. Paul's Cathedral was laid in 1679 and was completed 30 years later in 1709, on the architect's 76th birthday.

John Bunyan and Samuel Pepys

John Bunyan (1628-88), the English author, published "The Pilgrims Progress". Samuel Pepys (1633-1703), the English diarist, was appointed Secretary to the Navy in 1672. His diary is unrivalled for its intimacy and the human picture it presents of daily life in 17th century London.

Early Wet Docks

The two earliest wet docks in London were Brunswick Dock and Howland Great Wet Dock. Brunswick Dock at Blackwall, dating from 1660, was the first in Britain. Howland Dock of 1696-1703, on the south bank of the river between Rotherhithe and Deptford, became known as Greenland Dock from 1763 because London's Arctic whaling trade was based there. Their main function was for the fitting out and laying up of ships and the processing of whale carcasses.

The Great Fire of London 1666

18th Century Georgian London 1714-1836

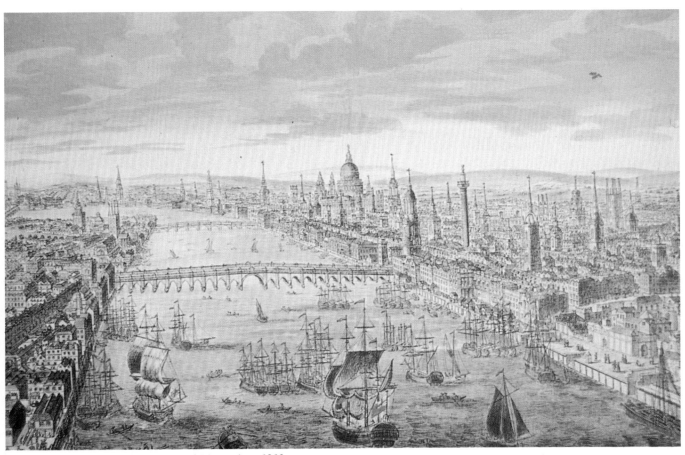

Rhinebeck watercolour of the City and River, circa 1810

The Georgian Period

For most of the Georgian period, England was at war with various countries. During the reigns of George I & II, the role of the Prime Minster began to emerge starting with Robert Walpole. Travelling outside London became dangerous because of attacks by highwaymen. Dick Turpin was just one who haunted the Hackney Marshes and Epping Forest! George I, acceded to the throne in 1714 precipitating the Jacobite rebellion of 1715. This attempt to overthrow the Hanoverian succession failed.

The Irish writer Jonathan Swift published his "Gullivers Travels" in 1726. Around 1738 saw John Wesley's conversion and the start of Methodism. In 1742 Sir Robert Walpole the British "Whig" Prime Minister resigned from the political scene after more than forty years. In 1745 Bonnie Prince Charlie led a Jacobite rebellion against George II - the Jacobites were defeated at Culloden. The great war with France started in 1793 and Wellington defeated Napoleon at the Battle of Waterloo in 1815, after which peace in Europe then prevailed. The accession of George IV took place in 1820 and in 1837 William IV died.

During the seventy years (1770-1840) when the Industrial Revolution was transforming England into the foremost manufacturing country of the world, canals were built to become the main arteries of inland communication and trade. Later the development of railways was closely connected with that of the coal industry. Watt's improved steam engine of 1765 came into general use after 1786 and was utilized to pull wagons along the tramways. In 1814 George Stephenson made the first practical steam locomotive "Puffing Billy", which you can see today in the Science Museum at South Kensington.

Arts and Architecture

The classical tradition was maintained by such master architects as Robert Adam and William Kent. Many gardens were planned and laid out by Capability Brown. The London Spas included Tunbridge, Epsom, Dulwich and Sydenham Wells. Joshua Reynolds was recognised as the master of English art. Other famous artists of the age included Turner, Gainsborough and Hogarth. Dr. Samuel Johnson (1709-84) produced the first English dictionary. Adam Smith published his "Wealth of Nations" and Edward Gibbon published his "Decline and Fall of the Roman Empire". The growth of Georgian London and its place as a leading city of the world was secure.

Australian Settlement

Following the American War of Independence in 1776, Britain was no longer able to send her prisoners to the American colonies. Another colony had to be found to avoid the outbreak of disease in the overcrowded prisons and places of confinement. In 1787 a decision was made to make a settlement at Botany Bay in Australia. The first fleet of eleven ships with 759 convicts left Portsmouth on 13 May 1787 and arrived in Sydney Cove on 13 January 1788. Of the eleven, two were warships, Sirius and HMS Supply, while the remainder were merchant vessels. A typical deportee was London thief William Cole, "indicted for feloniously stealing on the 2nd June last, one watch with gold case, a gold chain, three stone seals set in gold, all said to be the property of David Ross Esq!"

Overseas Trade Expansion

During the 18th century there was a considerable expansion in overseas trade and the construction of roads and canals stimulated the movement of goods within London and to other parts of the country. The profits gained in this expansion of trade and industry provided an abundance of commercial capital, while the developing banking system provided increased facilities for borrowing money to then use in industry and trade. There was growth in the power and influence of great trading companies such as the East India Company. During this period the trade of the Port of London quadrupled resulting in congestion and pressure of work on the wet docks.

Victorian and Edwardian London 1837-1913

Victorian London 1837-1901

In 1837 Queen Victoria came to the throne. About the same time the Royal Mail started and later the Penny Post was instituted. During 1844-5 the railway mania, with massive speculation and investment, lead to the building of 5000 miles of rail track. The Great Exhibition of 1851 was held at the Crystal Palace in Hyde Park. This exhibition celebrated the ascendancy of Great Britain in the market place of the World. The profits from this great exhibition were used to finance the building of the museums and colleges in South Kensington. The architect of the Crystal Palace was Joseph Paxton 1801-65. The building was revolutionary in its structural use of glass and iron. After the exhibition, the building was moved to South London and sadly caught fire in the mid 1930s and was destroyed.

The period 1850-60 saw the expansion of daily and Sunday newspapers. In 1859 Darwin published his "Origin of Species". Reforms to give male householders the vote were passed by Parliament in 1867. A great deal of poverty existed in London's East End. Matthew Arnold, British Poet 1822-88, described East Londoners as "those vast, miserable, unmanageable masses of sunken people." The Opening of the Suez Canal in 1869 gave Disraeli a controlling interest by buying canal shares. In 1882 Britain occupied Egypt.

Industrial Revolution

During the 19th Century engineering made great strides and saw the creation of a transport infrastructure, roads, railroads and water carriage were developed. Mid-century saw the development of the Industrial Revolution and economic growth increased. Engineers such as Brunel, McAdam, Telford, Rennie, Smeaton and Brindley played an important role in this Industrial evolution and advanced the supremacy of British engineering inventions and development. Huge prefabricated glasshouses were built of glass and iron, the best known being Burton's Palm House at Kew and Paxton's Crystal Palace. Brick warehouses were built along the Thames from the end of the 18th century and throughout the 19th century. Literary giants included Coleridge, Wordsworth, Byron, Shelley, Kingsley, Carlyle, Dickens, Kipling, Wells, Shaw, Yeats and Wilde.

The development of heavy industry, the railways, capital projects and the growth of the British Empire called for a sound business institution and London became the greatest money market in the world. At the same time the outward expansion of the suburbs resulted in a decline in the residential population of the City as the railways and other modes of transport made it possible for people to live at a distance from their work.

Edwardian London

Queen Victoria's death marked the end of an era. Edward VII's accession in 1901 saw the beginning of modern London as we know it today. It was a time of expanding commerce, with the formation of large public companies rivalling family firms; bustling department stores sold a wide range of luxury goods through new forms of advertising and marketing methods. A modern world which was made possible by electricity, the telephone, the excitement of motor car travel and the possibility of air flight. High society experienced the greatest period of luxury and entertained on a grand scale. A new cosmopolitan society was captured in glamorous portraits, and for the first time also through photographs and cinema news reels. At the same time massive demonstrations called for "Bread for our Children" and women campaigned for universal suffrage. The mass of people sought to create a more just and equitable society for working people. World War I started on 12th August 1914.

The Great Exhibition of 1851, held in Hyde Park

London Between World Wars 1914-1945

World War I and After

In June 1915, Londoners watched in horror as a German airship burst into flames in a ball of fire near the Tower of London. Later air raids by German Zeppelins were targeted at the docks. Londoners mostly went into their basements as the sirens were sounded. When the U-boat attacks were increased, cargoes were diverted to France. The greatest effect of the war was the loss of young men in the trenches during the battles in France and Belgium.

Following the Great War 1914-18, London saw many changes. In the 1920s steam ships finally replaced clippers, and lorries replaced the horse-drawn carts on the roads. Old Georgian, Victorian and Edwardian buildings were demolished making room for new modern architecture. The early period saw the introduction of the "Motion Picture" and the increase of splendid department stores. Large office blocks were constructed and West End squares were refurbished. New hotels were built and new roads constituted. Modern factories multiplied and a new industrial belt arose. By the mid-twenties, private housing took off in a spectacular fashion with the appearance on the scene of the Building Society.

Industrial Expansion

During the late 1920s and the early 1930s, the Greater London area actually doubled in size. Suburban properties spread out over the countryside and the population rose from around six million to over eight million in just twenty five years. London attracted the interest and capital of the emerging American millionaires. Heavy industrial investment saw large factory developments such as Ford Motor Company, General Motors, American Tobacco Company, Standard Oil, Trico, Hoover, Gillette, Firestone, Heinz, Colgate and Palmolive. Hollywood films completely dominated the cinema.

In 1922 the British Broadcasting Corporation (BBC) was formed. The Empire Exhibition to promote Britain's Imperial Trade was held at Wembley Stadium during 1925. In September 1929 the Wall Street Stock Market in New York collapsed causing the start of the worst recession in history. During the depression many of the old coal-based industries declined rapidly and large numbers of jobless people descended on London in search of work. In 1933 London Transport was formed. By 1935 about half Britain's large manufacturing companies had offices in the Capital. During 1936 King George V died, Edward VIII abdicated and the coronation of George VI took place in 1937. World War II started in September 1939.

London Blitz of World War II

On 7th September 1940, 400 planes bombed all parts of London. During this raid, high explosive and incendiary bombs caused great fires in the City. Air raids on London continued for 57 consecutive nights into early November. By the end of the war large tracts of the Capital had suffered considerable damage. Nearly one third of the City of London was lost in the blitz and the dockland areas in East London suffered badly during the air raids. Around the Eastern Dock at St. Katherine the Telford/Hardwick warehouses were destroyed and other warehouses in the vicinity were damaged, while at the Surrey Commercial Docks a high proportion of the transit shed accommodation was lost. Of the original range of 11 massive warehouses on the north quay of the West India Import dock only warehouses Nos.1 and 2 survived. In both world wars, cargo and passenger ships were adapted to operate as troopships, hospital ships and armed cruisers. In World War II particularly, losses of merchant ships were very heavy. The devastating effect on merchant shipping can be even more realistically understood when it is considered that between 1939 and 1945 the loss of shipping which normally used the Port of London amounted to four million tons.

Delivery vehicle 1925

Bombing of the Surrey Docks during the 1940 Blitz

Modern London 1945-2000

Festival of Britain 1951
In 1951 the Festival of Britain was organised to celebrate the development of arts, science and technology. The exhibition was held on a redeveloped South Bank site between County Hall and Waterloo Bridge. The whole site has since been developed into a huge arts complex with plain concrete buildings including the Royal Festival Hall, the Queen Elizabeth Hall, the Purcell Room, the Hayward Gallery and the National Theatre. During the summer the Centre is packed with visitors and school children for pleasure and educational purposes.

Queen Elizabeth II Coronation 1953
On 2 June 1953, over a million people lined the streets from Buckingham Palace to Westminster Abbey for the Coronation of Queen Elizabeth II. The procession was attended by leaders of the Commonwealth and other countries, riding in horse-drawn carriages and accompanied by the Queen's Guard and Troops. There were celebrations all over London that day.

Buildings and Urban Expansion
Post World War II many dwellings and office blocks had to be built. Structural techniques in steel and concrete incorporating slab, post and beam construction were developed further. The cantilevered framed building freed the interior of supporting columns and allowed lightweight external walls of glass or other light materials to be used for tower blocks. London overspill led to the creation of new towns such as Stevenage, Harlow, Hemel Hempstead, Crawley, Bracknell, and more recently Milton Keynes.

The Swinging 60s
During the 1960s London became one of the world's leading centres for music, fashion and design. People came from all over the world to buy goods from Oxford Street and Carnaby Street in the West End and King's Road, Chelsea. The streets were lined with new style boutiques selling the latest styles for the young. "Pop" music was at its golden age with such famous groups as The Beatles, The Rolling Stones and many others who dominated the music charts around the world. The Barbican Centre for arts and conferences, the largest of its kind in Western Europe, was opened in the City of London in 1962. Both the London Symphony Orchestra and the Royal Shakespeare Company reside here.

Closure of Upper Docks 1970s
The move downstream of much of London's port to Tilbury in the 1970s left a trail of empty warehouses and desolate wharves along the riverside. It also provided a new opportunity for London - to use the river and its environment and to enhance the character of the capital in all its aspects.

Commuter's Revolution of 1980s
Faced with the rapid increase in property prices in London after 1981 and the demand for a better quality of life, more living space and green fields, there was a huge upsurge in the number of people commuting to London. They sold their expensive London homes and moved to the surrounding counties where prices were still reasonable. Between 1982 and 1987 the number of commuters travelling by public transport increased from 770,000 to 925,000, while those travelling by private transport fell from 234,000 to 190,000.

London Towards 2000
Over the past decade London has been transformed and its skyline is now one of high rise buildings. In the City the Guildhall and St. Paul's Cathedral are surrounded by modern skyscrapers such as the National Westminster Tower, the Stock Exchange and Lloyd's. The Old Covent Garden Market has been converted into a fashionable shopping precinct while the fruit and vegetable trade has been relocated to new buildings at Nine Elms. The centuries old Billingsgate Fish Market was moved to the North Quay of the Isle of Dogs. London still remains the centre of Britain's financial and commercial power. Here too is the political heart of the nation, where laws are made in the Palace of Westminster and executed by members of Her Majesty's Government.

Centre Point, built 1966 at Tottenham Court Road

WESTMINSTER

Westminster, Whitehall and Trafalgar Square

Houses of Parliament and Big Ben Clock Tower

Houses of Parliament and Big Ben

Houses of Parliament
This magnificent Victorian Gothic style building designed by Sir Charles Barry took around 30 years to build between 1840-1868. It occupies the old site of a royal palace and hence is known as the Palace of Westminster. The oldest part is Westminster Hall which dates back to late 14th century. The Victoria Tower rises 102.5m above the River Thames and houses parliamentary archives. The Clock Tower, with its bell Big Ben, is a world famous landmark. Westminster Hall and the Houses of Parliament's Public Gallery are open for visitors at certain times. A flag flying from the Victoria Tower by day or a light in the Clock Tower at night signifies that Parliament is sitting. Westminster tube station faces the palace.

Westminster Hall
Built by William II (William Rufus), the Hall, 72m long, is the largest Norman hall in the world. At the end of the 14th century Richard II rebuilt the hall and replaced the roof structure with magnificent timber trusses. It is said that on New Year's Day in 1536 Henry VIII gave a great feast here for 6000 needy people!

House of Lords
The public entrance to the Houses of Parliament is in Old Palace Yard which leads to the Royal Staircase and the Royal Gallery, designed to provide a processional access for Kings and Queens at State openings. Beyond is the House of Lords with its beautiful interior containing the throne, brass railings and the candelabra. On the rear walls are large frescoes representing scenes from British history. Around the upper walls are eighteen bronze statues of the lords who made King John sign the Magna Carta.

House of Commons
From the House of Lords through Peter's Lobby you reach the House of Commons. The Chamber was destroyed by bombing in 1941 and was rebuilt to the same design and opened in October 1950.

George V & Oliver Cromwell Statues
By Thornycroft 1899, Cromwell's statue is sited in Old Palace Yard to the west of the Houses of Parliament. The King George V statue by Reid Dick 1947, is nearby.

Big Ben Clock Tower
The world famous Clock Tower of the Palace of Westminster is 97m high and 12m square in plan. Its bell, Big Ben, is a giant bell cast in East London at Whitechapel Bell Foundry in 1858. It takes its name from the stout Commissioner of Works, Sir Benjamin Hall, who was responsible for hanging it in 1859. The bell, 2m high and 2.75m diameter, weighs 13.5 tonnes and can be heard over a large area of Central London. The clock was built by Dent with the four faces each 7m in diameter. The minute and hour hands are 4.25m and 2.75m respectively. The clock striking the hour is a familiar sound in broadcasting.

Parliament Square Statues
There are four bronze statues: Sir Winston Churchill by Ivor Roberts-Jones 1973, Sir Robert Peel by Mathew Noble 1876, Field Marshall Jan Christian Smuts by Jacob Epstein 1958, and George Canning (Statesman) by Richard Westmacott. Opposite the Houses of Parliament, the statue of the famous American President, Abraham Lincoln, by Saint Gaudens 1920 is identical to the one in Chicago.

Boadicea
Also facing the Houses of Parliament, at the northern end of Westminster Bridge, is the famous statue by Thornycroft 1902, showing the Queen of the Iceni with her daughters riding a war chariot. She attacked London during the Roman period in 61AD.

Westminster Bridge
A fine Victorian bridge built 1854-56 to the design of the engineer Thomas Page in harmony with its historic surroundings. The seven cast iron arches rest on piers and abutments of grey granite. The handsome balustrades, lamp standards and spandrels in Gothic style match the Houses of Parliament on the north side of the bridge.

Westminster Abbey - 1000 Years History

Westminster Abbey
Adjacent to the Houses of Parliament is Westminster Abbey, one of the finest and oldest cathedrals on earth which houses the Coronation Chair. The Gothic style Abbey has been the place of crowning English Kings and Queens since Edward the Confessor who built the original church in 1065. Rebuilding was commenced by Henry III in 1245 and completed by Henry Yevele and others 1376-1506. Henry VII added the chapel early in 1503 with fine perpendiculars and magnificent fan vaulted ceilings. The towers were finished by Hawksmoor in 1734. Visit also the brass rubbing centre and the Undercroft Museum. The Abbey is extremely rich in metal work - mainly bronze and iron, mainly in grilles and surrounding tombs. During the Civil War, much of the Abbey was destroyed but replaced using the correct materials and styles relevant to the area and period from which they came.

There are tombs and memorials of past Kings and Queens and famous British subjects over the centuries. Among the royalty and statesmen buried here are Queen Mary and Queen Anne, Pitt the Elder and the Younger, Charles Fox who fought to abolish slavery, Sir Robert Peel, Lord Palmerston and Gladstone. Famous scientists and engineers who have memorials include Sir Isaac Newton, Lord Rutherford, Lord Kelvin, Michael Faraday, Robert Stephenson and Thomas Telford. Baroness Burdett-Coutts was the last person to be buried in the Abbey in 1906 and since then only ashes have been accepted. Westminster and St. James Tube stations are nearby.

Chapter House
On the east side of the Abbey is the splendid Chapter House completed 1245-55. It is entered through a richly carved double doorway. In the well lit octagonal chamber there are four tall windows (40 ft high by 20 ft wide) which fill the whole of the wall space. The tiles on the floor are original and date back to the 13th century. The first meeting of Henry III's Great Council was held here in March 1257, and between the middle of the 14th and 16th centuries it was used for the meetings of the House of Commons. Subsequently it was used to store public records until 1866. There are some fine paintings and sculptures.

Pyx Chamber
Built 1065-1090, the chamber houses the treasures of the Abbey reflecting its use as the strongroom of the Exchequer from the 14th to 19th centuries. It housed a pyx or box containing the standard pieces of gold and silver. The Queen's Remembrancer presides at the Trial of the Pyx in the Goldsmiths' Hall in the City of London when the coins of the kingdom are tested annually in an ancient ceremony.

Undercroft Museum
Many of the Abbey treasures are exhibited here including the wax effigies of sovereigns and their consorts which had been carried on top of the hearse in the funeral procession. It is believed that the 11th century room was once the Monks' Common Room.

Westminster School
Located on the south side of the Abbey's precinct building is the School built 1090-1100 and restored in 1959 following war damage. The College Hall Cloisters circa 1376 was the Abbot's state dining hall and is now used as the school refectory. The Palladian style College was completed by Lord Burlington circa1730. Former famous pupils include Ben Jonson, Sir Christopher Wren, Judge Jeffreys, Charles Wesley and Lord Russell.

St. Margaret's Church
To the north of Westminster Abbey, this 12th century church has beautiful glass windows and a variety of funeral monuments including Sir Walter Raleigh and Sir William Caxton, and the effigy of Blanche Parry who served as keeper of the royal jewels.

Methodist Central Hall
Built in Storey Gate 1905-11, the church is the principal Methodist centre with a hall having a capacity of 2,700. It is used for services, recitals, concerts and public meetings. The first assembly of the United Nations was held here in 1948.

Westminster Abbey

Westminster Cathedral and The Tate Gallery

Victoria and Pimlico
To the west of Westminster Abbey is Victoria which is situated at the centre of rail, coach and tube networks and is always busy with arrivals and departures. Further south is Pimlico which was developed mainly in the 1850's and is the home of the Tate Gallery.

Little Ben
A smaller version, 9m high, of Big Ben stands outside Victoria Station.

London Transport Headquarters
These offices built in 1920s near Victoria Station, are decorated with statues by Jacob Epstein, Henry Moore and Edward Gill.

Westminster Cathedral
In Francis Street, Victoria, is the most important Roman Catholic Church in England. Designed by Bentley and built 1895-1903 it is a wonderful Byzantine style Cathedral depicting early Christian architecture. With a 2 domed crossing and Italianate nave the campanile is 284 ft high with a cross at the top 11 ft in height. The interior of the church is covered with mosaic and opus sectile in Byzantine tradition and is ornamented with more than 100 different kinds of marble from quarries all over the world. The nave is the widest in England with eight columns of dark green marble. The campanile, which can be ascended by a lift, gives panoramic views of London.

Millbank
Millbank leads from the Houses of Parliament south to Vauxhall Bridge. First to catch the eye are the beautiful Victoria Gardens along the river. Nearby is Smith Square with its 18th century houses, conveniently situated for Members of Parliament. In the middle of the square is the English Baroque St. John's Church, designed by Thomas Archer and built 1713-28, and mentioned by Charles Dickens in his novel "Our Mutual Friend". The church was destroyed during the war and has since been restored as a music centre.

ICI Headquarters
Here is also the headquarters of the Imperial Chemical Industries (ICI), designed by Sir F Baines in 1928. The stone faced building has bronze doors with panels in high relief, representing man and his relationship with science and technology.

Vickers Tower
Situated in Millbank, SW1 and built 1960-63, the 34-storey tower has a special silhouette character due to its architectural plan form.

Tate Gallery
The gallery is in Millbank and near Pimlico tube station, and houses excellent collections of British and international paintings from the 16th century to the present day. It was designed by Sidney Smith and built in 1894 by donations from Henry Tate, the sugar millionaire, who also donated his private collection. The Gallery houses three main collections, the British Collection of paintings and sculptures from the 16th century to the present day, the 20th century Modern International Collection and the Turner Collection in the specially designed Clore Gallery opened in 1987. There are fine paintings by Turner, Blake, Hogarth, Nicholson, Spencer and Francis Bacon, and a rich collection of international paintings and sculptures since 1880, including paintings by Picasso, Chagall and sculpture by Epstein.

Westminster Cathedral

Map of Westminster, Whitehall and Trafalgar Square

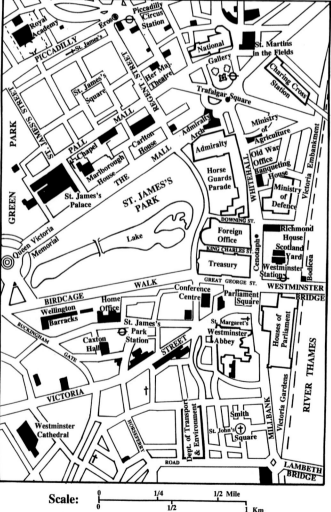

Scale:

Prime Minister's Residence and Whitehall

Trooping the Colour, Horse Guards Parade

Whitehall
The fine wide street leading from the Houses of Parliament to Trafalgar Square is used for state and ceremonial processions. It is lined with Government buildings and contains a number of notable statues.

Treasury Building
In 1845 Charles Barry completed this building on the site of an old Treasury started by William Kent in 1733. The interior was completely altered in 1960-64 when various remains of the Whitehall Palace of Henry VIII were discovered including the walls of the tennis courts. It houses the cabinet office and Privy Council.

Cabinet War Rooms and Museum
The underground emergency accommodation in King Charles Street SW1, was used by Sir Winston Churchill, the War Cabinet and the Chiefs of Staff during the Second World War and is open seven days a week. Visitors can view a complex of 21 historic rooms including the Cabinet Room and Map Room.

Foreign and Commonwealth Office
The building has fine architectural features including the Grand Staircase, the Oval Office, (formerly the Office of the Secretary of State for India), and the Foreign Secretary's Office. The offices were originally designed by Sir George Gilbert in the mid-nineteenth century. Since 1984 they have been refurbished to create a modern working space within the confines of the historic building.

The Cenotaph
This is the British Empire's memorial to "The Glorious Dead", the thousands of men and women who died during the world wars. It was designed by Sir Edward Lutyens in 1920.

Prime Minister's Residence
This house at 10 Downing Street has been the official residence of the Prime Minister since 1732 when George II offered the house to Sir Robert Walpole, Britain's first Prime Minister. Downing Street, the site of an old Roman Settlement, was built up by Sir George Downing, a Treasury Secretary, around 1680. The interior includes the Prime Minister's apartment, his private office and Cabinet Room, and the Secretary's Room used for civil servants. Its 250 year old staircase displays portraits of former Prime Ministers. It was rebuilt during 1960-64 but most of the historic features were retained. No.11 Downing Street is the home of the Chancellor of the Exchequer and No.12 of the Government Chief Whip.

Ministry of Defence
This large building with river frontage was completed in 1964 under which lies the wine cellar of Henry VIII, completely intact.

Richmond House
The façade of Richmond Terrace has been retained and the office buildings are modelled on the first Elizabethan era. The five most important rooms, recreated in their original form, are used as Minister's offices, committee and conference rooms.

Banqueting House
This beautiful House, built 1619-1622, to the design of Inigo Jones for James I, has magnificent ceiling paintings by Rubens in 1635 depicting the divine rights of Kings. For centuries the House was used for Court hospitality and entertainment and for a time it became a royal chapel. In 1890 Queen Victoria gave it to the Royal United Services Institute and it was renovated by the Government in 1963.

Horse Guards
The two horse mounted troopers of the Household Cavalry in Whitehall wear beautiful uniforms and are a great tourist attraction. The Ceremony of "Changing the Guard" currently takes place weekdays at 11am. The Queen's Guard is changed in front of Buckingham Palace at 11.30 am daily during the summer and every other day during the winter.

Horse Guards Parade
The fine Palladian building built 1750-58 to the design of William Kent is the headquarters of the Household Cavalry. Mounted troopers are posted daily from 10.am to 4.pm with relief every hour. The Trooping of the Colour takes place in the yard. Around the parade ground there are bronze statues of Kitchener, Wolsely and Roberts, and also a Turkish cannon made in 1524.

Paymaster General's Office
Past the Horse Guards is the Paymaster General's Office built by John Lane c 1732.

Admiralty Building
Designed by Thomas Ripley and built 1722-26, this building incorporates features from earlier buildings on the site. The Lords and Boards of The Admiralty charged with the provision, control and maintenance of the Royal Navy, occupy this building.

Trafalgar Square and The National Gallery

Trafalgar Square

The square was laid out by Sir Charles Barry in 1829. The 185 feet high Nelson's column is by William Railton, 1840. There are bronze lions by Landseer, 1868. The fountains are by Lutyens, 1948. The square is famous for its pigeons and as a place for political demonstrations. There is a terrace to the north side of the square with a broad flight of steps on either side leading down into the square. In each of the sloping east and west walls is a fountain of 1860. The Imperial Linear Measures of feet and inches are set against the north wall of the square. The column and fountain are lit by night as a tourist attraction in London.

Charles I and Nelson Statues

The 16 feet high Nelson statue by Bailey, 1843, is on top of the column in Trafalgar Square. In front of Nelson's Column is a statue of Charles I facing towards Whitehall. By Le Sueur 1633, it is the oldest equestrian statue in London and one of the finest. Ordered to be destroyed during the Civil War and hidden until the Restoration, it was erected on the present site around 1675. On either side of the column are bronze statues of Generals Napier and Havelock.

Other Landmarks

The east and west sides of the square are flanked by plane trees. On the north side is the National Gallery. Other buildings include the Church of St. Martin-in-the-Fields, Canada House designed by Sir Robert Smirke and converted in 1925, South Africa House by Herbert Baber 1935, and Australia House with its bronze Horses of the Sun over the entrance by Bertram Mackennal 1919. On the corner of Strand and Charing Cross Road is a plaque marking the central point of London from which all distances on sign posts are measured.

Christmas Tree from Norway

Every year at Christmas a huge spruce tree is erected as a gift from the Norwegian people in recognition of their gratitude to the British people for their help towards Norway during the Second World War. On New Year's Eve, the square is the scene of merry celebrations.

Admiralty Arch

To the south of the square is the massive Edwardian triple arch which was designed by Sir Aston Webb and erected in 1911 in memory of Queen Victoria. It forms the entrance to the Mall from Trafalgar Square.

National Gallery

Located on the north side of Trafalgar Square, this gallery is one of the finest in the world and its exhibitions display the major works of the greatest painters in history and some of the world's most renowned works of art. With over 2000 paintings, they include treasured collections of great Italian, French, Venetian, German, Dutch and Flemish painters. The imposing building dates back to 1838 but has been extended several times and mostly rebuilt in recent years. Charing Cross and Trafalgar stations are close. Special exhibitions are held regularly.

James II Statue

A fine statue by Gibbons 1686, located outside the National Gallery. It shows the King in Roman costume.

George Washington Statue

Presented by the State of Virginia, USA 1921, it is a replica of the one in Richmond, the statue is outside the National Gallery.

National Portrait Gallery

Adjacent to the National Gallery and founded in 1856, this excellent gallery displays individual portraits of distinguished British men and women of the past.

St. Martin-in-the-Fields Church

Situated to the east of Trafalgar Square, this fine church by James Gibbs, was built in 1721 on the site of an old church erected by Henry VIII in 1544 during a period of plague in London. The first church service broadcast by the BBC in 1924 came from here. George I donated the organ. The painter Constable was married in the church and Francis Bacon and Chippendale are buried in the old churchyard. After the first World War, the crypt was opened for homeless soldiers returning from France, and during the second World War it was used as an air raid shelter. It is now the home of a fine orchestra: the Academy of St Martin in the Fields. The church is well known for its welfare work.

Trafalgar Square and National Gallery

Buckingham Palace and St. James's Palace

Buckingham Palace

The Mall and Pall Mall

The Mall is a fine tree-lined street leading from Trafalgar Square to Buckingham Palace and its surroundings demonstrate the affluence and wealth of early 19th century London. The many elegant houses in nearby Pall Mall include two buildings by Sir Charles Barry, the Italian Renaissance Travellers Club built 1829-32 and the Reform Club, 1837-41.

Buckingham Palace

In 1837 Queen Victoria made Buckingham palace her permanent London home and every reigning Monarch since has done the same. It is named after the Duke of Buckingham who built the fine mansion in 1703. George II acquired the property in 1762 and his son George IV commissioned John Nash in 1825 to remodel it as a palace. The present day quadrangle layout was formed in 1847 and the facade was refaced by Sir Aston Webb in 1913 in Portland stone. Prince Charles was born here on 14th November 1948. The State Ballroom and its Royal Dais are used for investitures and state banquets. The west wing faces 40 acres of beautiful gardens where Royal garden parties are held. The Royal Standard is flown above the palace when the Queen is in residence. Her Majesty the Queen has opened the doors of Buckingham Palace for the public to visit. The Picture Gallery and the Green Drawing Room have magnificent chandeliers, furniture and paintings. The decorations and curtains are mainly in crimson and gold leaf colours.

Queen Victoria Memorial

In front of Buckingham Palace, the impressive memorial by Brock 1911 contains a fine figure of Queen Victoria.

The Queen's Gallery

Opened in 1962, the gallery displays a selection of art treasures from the Royal Collection since the days of Charles I. The exhibitions include paintings by famous artists, drawings, furniture and other priceless items. Built originally as a conservatory for Buckingham Palace by John Nash, the gallery was converted into a Chapel in 1843.

Royal Mews

Adjacent to Buckingham Palace, the Mews houses the Queen's horses and Royal carriages, including the magnificent Coronation Coach.

Clarence House

This was the House of the present Queen Elizabeth and the Duke of Edinburgh until her accession to the throne in 1953. Built by William IV it was later occupied by Queen Victoria's mother. Near Buckingham Palace and overlooking Green Park it is now the London residence of the Queen Mother.

St. James's Palace

The Palace is the headquarters of the Yeomen of the Guard and the Lord Chamberlain. Construction of the Palace was begun by Henry VIII in 1532 and it became the official London residence of Royal Monarchs in 1698 when Whitehall Palace was destroyed by fire. There are four courts: Ambassadors' Court, Friary Court, Engine Court and Colour Court. The last court is approached through the original gatehouse with its flanking octagonal turrets, standing at the southern end of the building. The balcony overlooking the Friary Court is the place where the Royal proclamation is made: "The King is Dead. Long live the King."

Marlborough House

Off Pall Mall in Marlborough Gate the house designed by Wren 1710 contains a fine painted ceiling by Gentileschi that was originally designed for the Queen's House at Greenwich. The Queen's Chapel in the grounds is by another famous architect Inigo Jones 1626.

Carlton House Terrace

Built by John Nash 1827-32, these two magnificent terraces at the eastern end of the Mall are "Grace and Favour" houses, leased at peppercorn rent to notable persons such as senior army officers and Civil Servants for past services to the Crown.

Coronation of Her Majesty Queen Elizabeth II

The Queen's Coronation

The Coronation of Her Majesty Queen Elizabeth II took place on Tuesday, 2 June, 1953. As the mother of two young children the Queen had already been on the throne some 16 months when she was crowned. She was just 25 when her father the King died. Both Prince Charles, four years of age and Princess Anne three, attended the Coronation in the company of the Queen Mother and the Queen's sister, Princess Margaret. Almost a year was spent planning and rehearsing for the ceremony. Prince Philip chaired the organising Coronation Committee. As each day passed, the Palace bustled with activity and preparation. The royal collection of gold plate, cutlery, trays, candlesticks and vases were cleaned and polished. The State Coach, made for George III and used for every coronation since that of George IV, was refurbished. The old iron-shod wheels were replaced by solid rubber tyres to cut down the jolting. Concealed lighting powered by batteries was installed under the seats so that the Queen was clearly seen as she rode through London.

Coronation Route

The map shows the extended route requested by the Queen so that as many people as possible would see the royal procession. When she left the Abbey, the Queen held the Orb in one hand and the Sceptre in the other. Inside the coach a small shelf took the weight of the Orb and a ring helped support the Sceptre.

Crowning at Westminster Abbey

At Westminster Abbey the Queen was met by the Archbishops of Canterbury and York. At the altar inside the cathedral, she knelt with her right hand on the Bible, and took the oath of monarchy. Her maids of honour removed her robe and jewellery before she sat in the Chair of King Edward. Four Knights of the Garter screened her with a cloth of gold as the Archbishop of Canterbury anointed her hands, breast and head with the consecrated olive oil. The ruby and sapphire ring was placed on the fourth finger of her right hand and she was handed the Royal Sceptre and the Orb. The Crown of St Edward was then dedicated and lowered onto the Queen's head to the sound of trumpets outside and guns firing at the Tower of London. The Queen was then moved to her throne. Prince Philip, was the first to kneel down and pay her homage, followed by other Dukes, Marquesses, Earls, Viscounts and Barons. The ceremony completed, the Queen left the Abbey for the long procession to Buckingham Palace. There, the Queen appeared on the balcony with her family to acknowledge the cheers of the great crowd outside.

Her Majesty Queen Elizabeth II

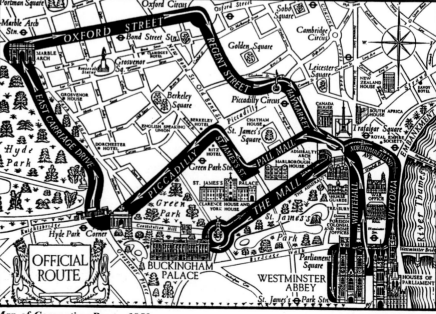

Map of Coronation Route, 1953

Hyde Park and Harrods of Knightsbridge

Hyde Park
This is the largest of the Royal Parks with over 300 acres of green fields, trees and other attractions. The adjoining Kensington Gardens extends to 275 acres. The manor belonged to Westminster Abbey but was taken over by Henry VIII in 1536 and converted into a royal deer park. The Stuart Kings used it for horse racing and ultimately in 1637 it became a Royal Park open for the public. In 1851 the huge Crystal Palace was erected here for the Great Exhibition. Afterwards it was re-erected at Sydenham Hill, South London, but was destroyed by fire in 1936. The Serpentine was added by George II. In 1872 Speakers' Corner was established at the eastern end of the Park near Marble Arch, as a place of unbridled free speech. There are a number of monuments and statues to see. The park is an ideal centre for green walks.

Hyde Park Corner
Hyde Park Corner consists of three classical style triumphal stone arches at its entrance. They were built in 1825 by Decimus Burton.

Constitution Arch
The Arch was built in 1828 as a northern gate to the Buckingham Palace estate. The magnificent imperial bronze chariot carrying Victory replaced the statue of the Duke of Wellington in 1912.

Royal Artillery Memorial
A fine war memorial at Hyde Park Corner erected by Jagger in 1925. The great howitzer is aimed at the Somme where the men it commemorates, lost their lives during World War I. The bronze figures of soldiers are excellent.

Wellington Statue
The equestrian statue of the Duke and four of his soldiers erected in 1888 by Boehm. The Duke rides his favourite horse 'Copenhagen' and he looks towards Apsley House where he lived.

Apsley House and Wellington Museum
Overlooking Hyde Park Corner and Green Park, the building was designed by Robert Adam and built 1771-78 and was later altered by the Duke of Wellington. The museum contains paintings, silverware, porcelain and military relics (071-499 4676).

Belgravia and Knightsbridge
South of Hyde Park Corner, this area consists of 19th century Regency squares and mews designed by Thomas Cubitt from 1825 onwards. The fine estate includes Belgrave, Chester and Eaton Squares. The Regency style squares with their gardens are connected by wide streets in a manner similar to that of Nash for Regents Park. On each side of the squares there is a range of stuccoed houses, decorated with raised porticoes, columns, pilasters and pediments. Belgrave Square is one of the most attractive in London.

Knightsbridge like many other parts of London was a small hamlet through which the River Westbourne flowed. Legend has it that two knights were once locked in mortal combat on the bridge over this river. Near this site is Harrods department store.

Harrods of Knightsbridge
This superb boutique store and centre of excellent fashion with a vast selection of high quality goods, is situated next to Knightsbridge tube station. Founded by Henry Charles Harrod in 1849 in smaller premises, it now occupies a large building in dark terracotta bricks with its own livery, designed by Stevens and Munt and completed in 1905. There are fabrics from France, Italy, Germany and Switzerland. In the magnificent marbled Edwardian food halls, the walls are covered with beautiful Art Nouveau tiles depicting the culinary delights displayed there. The needlecraft department has tapestry, wool and silk threads. Large stocks of handbags and leather goods with unusual colours and designs. The bookshop has a fine selection of books and magazines. A variety of services are offered to customers. The annual Harrods Sales are renowned throughout the world.

Currently owned by Mohammed Al Fayed, Harrods has over 300 departments in 20 acres of selling space. About 35,000 customers visit the store each day during the January sales. The fine new Jewellery Room houses seven world famous jeweller's shops. The Georgian Restaurant is Harrod's premier eating place on the fourth floor, where you may enjoy traditional English food while listening to a resident pianist. Enjoy your visit!

Harvey Nichols
In London's fashionable Knightsbridge at the corner of Sloane Street, this was founded by Benjamin Harvey in 1813, as a Linen Store. Colonel Nichols became a partner in 1820 and the store expanded into oriental carpets, silks and fabrics. The store's contemporary environment comprises of seven floors of ladieswear, cosmetics, fashion accessories, menswear, childrenswear, fashion for the home and food. There are exclusive boutiques of famous collections and a restaurant on the fifth floor.

Harrods of Knightsbridge

Victoria & Albert, and Natural History Museums

Kensington, Chelsea and Fulham

South Kensington has a group of famous museums and buildings honouring the development of arts, science and technology, they include the Victoria & Albert Museum, Royal Albert Hall and Science Museum. Kensington Palace was rebuilt by Christopher Wren for William III in 1689. It stands in the beautiful grounds of Kensington Gardens adjoining Hyde Park. Kensington High Street is a fashionable shopping area of Central London. Chelsea has an equally famous shopping area in Kings Road, which was built as a private coach road for Charles II. The boutiques and shops for the latest fashions intermingle with cafes, pubs and restaurants. Parallel to King's Road is Fulham Road which also has its boutiques, restaurants and wine bars. Fulham Pottery started in 1671 is still functioning. Fulham Palace with its beautiful herb garden, until recently the residence of the Bishops of London, is one of the oldest buildings in the capital and is open to the public.

Victoria and Albert Museum

The superb museum in Brompton Road was designed by Sir Aston Webb, the foundation stone was laid by Queen Victoria in 1899 and the building was opened by Edward VII in 1909. The national museum of applied arts contains collections of all kinds and periods.

There are important exhibitions of paintings, sculpture, calligraphy, armour and weapons, carpets, clocks, costumes, fabrics, embroideries, engravings, ironworks, jewellery, pottery, tapestries, watercolours and woodwork. The collections illustrate the Medieval, Gothic and Renaissance Arts, Islamic, Indian and Far Eastern Arts, English furniture and decorative arts. The paintings include those of Constable and Turner. A new extension was opened in 1982 for temporary exhibitions including design of objects that are part of our daily life.

Brompton Oratory Church

Adjacent to the east of the Victoria and Albert Museum, this Roman Catholic Church is in Italian Renaissance style; the interior is richly decorated in carvings and mosaics.

Natural History Museum

Situated in Cromwell Road, on the west side of the Victoria and Albert Museum, this magnificent Gothic style building by Alfred Waterhouse was built in 1881. The façade has horizontal bands of buff and sombre blue terracotta. The arched entrance is flanked on either side by towers 192 feet high. This museum houses the national collections of botany, palaeontology, zoology and entomology. The 90 foot model of the blue whale and the large model of a dinosaur are great attractions. The bird gallery is excellent. The walls have carvings of animals and plants. The new annexe is well integrated with the rest of the building. The museum is ideal for the instruction of school children.

Islamic Cultural Centre

Located on an island site in Cromwell Road, opposite the Victoria and Albert Museum, the Ismaili Centre was built in 1983 as a religious and cultural centre for the Ismaili Muslim Community. The two storey building has a roof garden. The modern design incorporates graphic motifs which have been worked into the granite facing.

Baden-Powell House

Opposite the Natural History Museum in Queen's Gate is the Scout Association Headquarters with an international hostel containing the records of the movement and mementos of Lord Baden-Powell.

Air and Coach Terminals

Further west along Brompton Road, there was the West London Air Terminal, the main London terminal for Heathrow Airport. Nearby is the National Coach Terminal which is used annually by thousands of UK and foreign travellers. These services cover the entire country.

Museums in South Kensington

The Science Museum and Geological Museum

Science Museum

Along Exhibition Road is the outstanding Science Museum designed by Sir Richard Allison 1914, which tells the story of science applied in industry, with practical models, steam engines, motor cars and various aspects of applied physics and chemistry. The exhibits include personal relics of the great scientists and engineers: beam engines by Thomas Newcomen 1771 and James Watt 1788; the oldest locomotive in the world 'Puffing Billy' of 1813; George Stephenson's "Rocket" of 1828. There are two beautiful cars; a Mercedes Benz 1888, and a Rolls Royce dated 1904, both in excellent working order.

In the entrance hall are the busts of Isaac Newton and Albert Einstein. In the basement there is a full-scale replica of a working mine and a gallery of dioramas on the development of transport specially intended for children. Old workshop machinery and clocks driven by weights, including one dating back to 1392, are exhibited on the first floor. There is also a lecture theatre with topics including astronomy. The ships' gallery on the second floor contains models of everything from small fishing boats to steam powered warships. There is a full scale model of a ship's bridge with radar equipment. Technological developments are explained in the marine engineering and navigation sections. Models of 18th to 20th century ships are exhibited in the Docks and Diving gallery.

The third floor is devoted to electricity and magnetism and contains the Royal collection of scientific instruments originally belonging to George III. There is also a gallery of aeronautics and early aeroplanes. The priceless collections contained within the museum are very instructive for children and adults alike. There are interesting features on space travel including a launch pad, space exhibition and Lunar module. The extensive museum library is invaluable to the research engineers and scientists of the neighbouring Imperial College.

Geological Museum

The museum was founded in 1837 and was based in Jermyn Street. The present building, by Sir Richard Allison next to the Science Museum, was completed in 1933 and houses one of the largest collections of gems, precious stones and fossils in the world. It deals with the earth's physical and economic geology and mineralogy. The ground floor exhibitions tell the story of the universe. The entrance hall contains cabinets full of diamonds, rubies, emeralds and sapphires. The first floor is devoted to the geology of Great Britain with a huge selection of fossils.

The Science Museum's East Hall

Map of Hyde Park, South Kensington and Kensington Palace

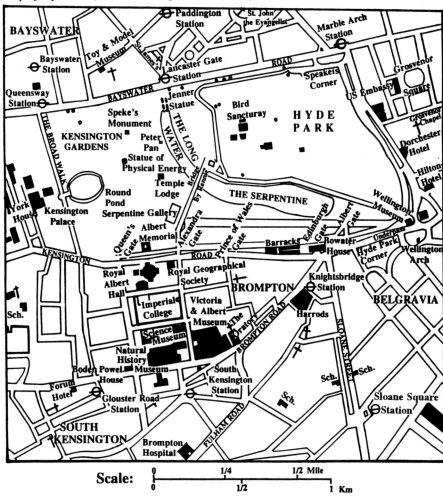

There is a piece of the Moon, collected by the Apollo astronauts in 1972. Samples of the stones and materials used for the buildings of London are on display on the second floor. The reference library is open to the public.

Imperial College

The college for science and technology was established in 1907 for advanced training and research applied especially to industry. It is part of London University.

Royal Albert Hall and Kensington Palace

Royal Albert Hall

Opened in 1871 by Queen Victoria the Hall was dedicated to the memory of her husband, Prince Albert. It faces the Albert Memorial in Hyde Park. Originally called the Hall of Arts and Sciences the red brick hall, designed by Francis Fawke in the style of a Roman amphitheatre, is 400 metres in circumference with a huge glass and steel roof. Heavily ornamented with terracotta, the frieze on the upper part illustrates the Triumphs of Arts and Sciences. The beautiful interior accommodates 8000 people and comprises the Arena, the Amphitheatre Stalls, three tiers of Boxes and the Gallery Promenade. The organ contains 10,000 pipes and is one of the largest in the world. The Hall is the venue of pop and classical concerts, conferences, indoor sports and rallies. Since 1941 the famous Henry Wood Promenade Concerts have been held here every summer. South Kensington tube station is within easy walking distance.

The Albert Memorial

The Memorial was erected in memory of Queen Victoria's husband. When Albert died in 1861, Victoria was inconsolable and withdrew from public life for much of the rest of her reign. In 1876, the Albert Memorial, a sculpture of metal, mosaic, gilt and enamel was erected in Hyde Park to his memory. The fifteen foot statue of Albert holds a Great Exhibition catalogue of 1851, while in the tall spires are statues of those virtues so dear to the Victorians - Faith, Hope, Charity, Chastity and Temperance.

Serpentine Gallery

In a pleasant location within Kensington Gardens near Albert Memorial, the art gallery provides a platform for young artists during the summer and for more established artists at other times.

Royal Geographical Society

Founded at Kensington Gore in 1830, the Society has made considerable contributions to the advancement of geography and organization of expeditions into the unknown parts of the world. Past explorers include Livingstone, Scott and Hillary. The map room has a collection of both old and new maps and is open to the public. The Statue of David Livingstone by J B Huxley-Jones was erected in 1953.

Royal Kensington Palace

In 1689 William III bought Nottingham House from the Earl of Nottingham and commissioned Sir Christopher Wren to upgrade it to a palace. Subsequently George I extended the palace to the design of William Kent. It remained the main royal residence until George II died in 1760. Queen Victoria was born here in 1819 and lived here until she became Queen in 1837 when she moved to Buckingham Palace. The State Apartments are open to visitors but the private apartments are occupied by some members of the Royal Family. The Apartments include Queen Victoria's Nursery. The Court Dress Collection, also open to the public, tells the story of fashions worn by ladies and gentlemen to the Royal Court, from 1750 onwards (071 937 9561). Kensington Gardens had a road from the Palace to the West End; part of this survives as Rotten Row in Hyde Park. The 275 acre park was opened to the public in 1843 and a Flower Walk created. The statue of Queen Victoria was erected in 1893 and that of William III in 1907. The oak and beech trees along Broad Walk were planted in 1953.

Peter Pan

A superb statue by George Frampton 1912, of the fairy tale character Peter Pan, was erected for children in Kensington Gardens.

Royal College of Art

At Kensington Gore this college of fine art was established in 1837, and in 1967 by Royal Charter became a university institution for awarding degrees. It has a high reputation nationally and internationally.

Royal Albert Hall

Commonwealth Institute, Olympia & Earls Court

Earls Court Exhibition Hall

Commonwealth Institute

Inaugurated as the Imperial Institute by Queen Victoria on 10 May 1893, the institute is the cultural and educational centre of the 50 Commonwealth countries scattered throughout the world. It plays an important role in raising the public awareness of the people of the Commonwealth to their homelands and topical issues. There are permanent country exhibitions, conferences, and a variety of programmes and activities. High Street Kensington tube station is within short distance along the High Street.

Leighton House Museum

Near the Commonwealth Institute in Holland Park Road is the home of Frederick Leighton, the classical 19th century painter and President of the Royal Academy. Built between 1864 and 1866 to the design of the architect George Aitchison, it has richly decorated rooms and excellent woodwork. The centrepiece of the house is the Arab Hall, added 1877-79, to house a collection of Islamic tiles, one of the finest in Europe. The house also contains a fine collection of Victorian aesthetic art including a collection of paintings by Lord Leighton.

Holland House and Park

The Palace of the Baron Holland built 1605-40, was damaged during the second world war and was subsequently restored. It is one of the finest Jacobean buildings remaining in Central London. The East Wing has curved Dutch-style gables. The Gate Piers are by Nicholas Stone to the design of Inigo Jones (1629). Holland Park is part of the original estate and has a charming Dutch Garden. In 1991 a new Japanese Garden was opened by HRH Prince Charles and the Crown Prince of Japan. The 'Wilderness' on the north side of the park is the largest area of natural woodland in Central London.

Antique Market

Situated in High Street Kensington, the market contains a large number of antique dealers who will refund the full amount if a customer has been unwittingly sold a fake article or work of art.

Portobello Road Antique Market

London's leading and largest Saturday Antique Market is situated in Portobello Road near Notting Hill Gate tube station. The Market starts at about 5.30 am, mainly with professional dealers trading amongst themselves. By 10 am the street is full of visitors, buyers and traders. Some dealers finish at lunch time, but many stay until 5.00 pm. Collectors and the general public flock to the area, roaming around the stalls and shops looking for bargains and reasonably priced antiques. Visitors and tourists come from all over the world. The galleries and stall holders sell books, jewellery, clocks, ceramics, toys and all types of art objects.

Earls Court Exhibition Hall

The large reinforced concrete building opposite Earls Court tube station covers an area of 12 acres. The 250 ft. wide hall is used for public exhibitions including the Royal Tournament, the Royal Smithfield Show and the Boat Show.

Olympia Exhibition Hall

A short walk from High Street, Kensington, the hall was opened in 1884 and has held a variety of exhibitions. The first Motor Show was held in 1905 and the first Horse Show 1907. Regular annual events include the Daily Mail Ideal Home Exhibition and Cruft's Dog Show (see page 121).

Chelsea Royal Hospital and The Army Museum

Chelsea Royal Hospital

The Royal Hospital was founded by Charles II in 1682 as a home for old and disabled soldiers. There are about 400 residents, who wear colourful coats, scarlet in summer and blue in winter. The Hall contains many flags and trophies and the Chapel is beautifully decorated. It was designed by Wren and later additions were made by Robert Adam and Sir John Soane. Part of the hospital was destroyed during the Second World War and a new infirmary was opened by the Queen Mother on 22 February 1961. More recently a library and a Roman Catholic Chapel have been constructed. The hospital grounds cover an area of 66 acres including the site of the Royal Horticultural Society's annual Chelsea Flower Show and extend to the Embankment along the Thames. The small museum contains a collection of relics representing over 300 years of the Hospital's history. The nearest tube station is Sloane Square.

National Army Museum

Within a short walk from Sloane Square tube station, the museum is adjacent to the Royal Chelsea Hospital. It tells the history of the British armies from the Yeomen of the Guard of 1485 to the soldiers of today. Exhibitions show how soldiers lived, worked and fought through five centuries. There is a large model of the Battle of Waterloo with 70,000 model soldiers. You can also see portraits by Reynolds and Gainsborough, the lamp used by Florence Nightingale and the skeleton of Napoleon's horse, Marengo!

Chelsea Physic Garden

Founded by the Society of Apothecaries in 1673 to grow plants for medicinal research, the gardens contain a valuable collection of rare herbs and medicinal plants. It is near Chelsea Royal Hospital (071-352 5646).

Crosby Hall

Crosby Hall was built in the 15th century by Sir John Crosby (Lord Mayor of London) as a merchant's palace in Bishopsgate in the City of London, and featured as 'Crosbies Place' in Shakespeare's King Richard III. Early in the 1900s it was bought by the Chartered Bank of India, Australia and China for demolition and a site for their new head office. The Bank stored the materials and the London County Council with financial help from the City of London Corporation re-erected the hall on its present site in Danvers Street in Chelsea in 1908. The interior of the building with its arched oak roof, carved pendants and vaulted oriel is splendid.

All Saints Church

This small church was rebuilt after being bomb damaged in 1941 and retains its 13th

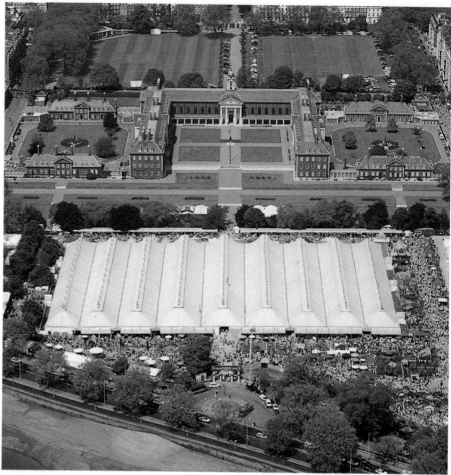

Chelsea Royal Hospital and Flower Show

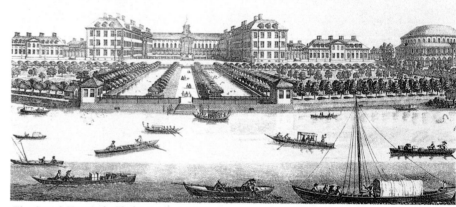

The Royal Hospital, Chelsea, circa 1751

century chancel and its name from the 15th century. Sir Thomas More, who built a country house nearby, attended this church. The famous one time Chancellor of England refused to accept Henry VIII as Head of the Church and was executed at the Tower of London.

Kings Road, Chelsea

Chelsea has a famous shopping centre at Kings Road which was built as a private coach road for Charles II. Like Carnaby Street in Soho, it has a range of boutiques with high quality clothes for young people.

Chelsea Bridge

This modern steel suspension bridge of 1937 replaced an older iron suspension bridge. The architects were Topham Forrest and E P Wheeler; the design was by the Consulting Engineers Rendel, Palmer and Tritton.

Carlyle's House

The writer Thomas Carlyle lived in this 18th century house at 24 Cheyne Road, Chelsea for 47 years until his death in 1881. Charles Dickens, Robert Browning and Lord Tennyson were visitors to the house. Displays include Carlyle's desk (071 352 7087).

Chelsea Harbour and Historic Fulham Palace

Chelsea Harbour

This is undoubtedly an innovative new development along the River Thames close to Wandsworth Bridge. Previously, it has been a railway yard and a canal basin where coal was trans-shipped for power stations. It combines a luxury complex in eighteen acres set around 75 berth yacht mooring on the banks of the Thames. There are flats, townhouses, offices, shops, restaurants, gardens, a hotel and leisure facilities. From it you can see the beautiful sweep of the river. It has its own hotel and marina and is a stop for the Riverbus on its way to Westminster and Docklands. A number of restaurants operate in the area including Chantegrill, Memories of China and The Waterfront. The diners have free use of the underground car park serving the development. For a more romantic arrival many diners will obviously prefer to travel on a speedy Riverbus. To land at the River Pier, stroll around the marina and to arrive at the front door of the restaurant of one's choice is a nice experience.

Wandsworth Bridge

This modern bridge of 1940 is constructed of huge steel cantilever beams pinned down on the banks and resting on two narrow streamlined piers, widely spaced. The engineer was Sir Pierson Frank and the architects were E Wheeler and F Hiorns.

Putney Bridge

Designed by Sir Joseph Bazalgette and constructed 1882-1886 of granite with five segmental arches and rusticated voussoirs, its handsome cast iron lamp standards surmount the plain parapet in the centre of each arch. The first Putney Bridge was built of timber in 1729 by the King's Carpenter, to replace an old ferry.

Western Pumping Station

Designed by Sir Joseph Bazalgette as part of the London drainage systems, it was built 1872-5 by the Metropolitan Board of Works on the north bank of the Thames in Chelsea. The building is of Italian style and is listed by the Department of Environment.

Fulham Palace

Screened from the River Thames by trees and a delightful garden of herbs, this historic residence of the Bishops of London until 1973, is a relaxing and peaceful place to visit. It is Fulham's most famous historic monument. The Grade I listed building dating from 1704 has a Tudor courtyard and a Georgian east front. The beautiful courtyard is c1520 but the bell tower dates from the 18th century. Visitors can enjoy the botanic beds, herb garden and wisteria pergola within the old walled kitchen garden. A Museum is contained in the early 19th century Palace in two major historic rooms of Bishop Howley's dining room and the Porteus Library. Here you can see the Palace's paintings, stained glass, a Bishop's cope, and find out about the Palace's history. Evidence of early Neolithic and Roman settlement lies under the east lawn. The site is protected as a Scheduled Ancient Monument. Located in Bishop's Avenue off Fulham Palace Road, SW6, Putney Bridge tube station is within 10 minutes walk.

Chelsea Harbour and marina

THE WEST END

West End theatres

Victoria Embankment and the London Savoy

Victoria Embankment

The Victoria Embankment extends from Westminster Bridge to Blackfriars Bridge along the north bank of the Thames. The river wall and cast iron lamp standards were built 1860-70 to the design of the Metropolitan Works Chief Engineer, Sir Joseph Bazalgette as part of the construction of the northern outfall underneath. The retaining wall and parapet, piers and stairs are in granite. Standing on dies in the parapet are elaborate cast iron lamp standards with intertwined dolphins and globe lanterns.

Memorials and Statues

Along the Embankment the Royal Air Force Memorial is a modern 20th century gilt eagle standing on a Portland stone pylon, designed by Sir R. Blomfield. The bronze group of Boadicea Statuary, was erected in 1902 by Sir John Isaac Thorneycroft. The Sir Joseph Bazalgette Memorial was erected in 1899 by George Simmonds. The bronze bust of the engineer in a circular niche, is enclosed in a large marble wall plaque. The Sir William Gilbert Memorial by Sir George Frampton unveiled in 1913, is another bust set in the niche. There are two other bronze plaques by Frampton.

Cleopatra's Needle

The oldest statue in Britain, this pink granite Egyptian obelisk circa 1500 BC stands on a granite pedestal flanked by bronze sphinxes, about halfway along the embankment. It was brought to London in a specially designed vessel and erected in this position in 1878. The Victorian sphinxes were designed by G Vulliamy and made by C Mabey.

The Queen Mary

In 1989, the Queen Mary, a converted Clyde pleasure steamer, 260 feet long and built in 1933, was opened as a new floating £2.6 million bar/restaurant complex on the Victoria Embankment. It is a welcome addition to the Thames Tourist amenities and is moored just west of Waterloo Bridge.

Charing Cross Embankment Complex

The Charing Cross redevelopment along the Embankment comprises of three major components: a new office building above the railway tracks and platform; an infill building on the west side of Villiers Street providing an extension to the station concourse at the middle level, and retail and leisure facilities on the lower levels; the conversion and renovation of the vaults below the station, to provide mixed retail and catering uses, includes the rebuilding of the premises for a 250-seat Players Theatre. The barrel-vaulted building has a bravura reminiscent of the great railway age.

Savoy Chapel and Hotel

Famous for the Gilbert and Sullivan operas in the adjacent theatre, this riverside world famous hotel between the Strand and the Embankment has an exciting past stretching back to the Middle Ages. During the 13th century, the Count of Savoy owned a manor house here which later became the old Savoy Palace. The sole survivor from the earlier days is the peaceful Chapel of Savoy which was gutted by fire in 1864 and restored by the architect Sidney Smirke and paid for by Queen Victoria. The entrance is modern, built in 1958, when a robing room and chaplain's vestry were added.

Savoy Theatre

In 1881 the design of the present Theatre was prepared by C J Phipps following the purchase of the land by Richard D'Oyly Carte. When opened with the opera Patience, the theatre was the first in the world to use electricity for lighting. To allay fears of fire, an electric bulb was smashed on the stage to prove that there was no risk! Subsequently, the hotel opened in 1889 offering wealthy guests magnificent views over the Thames. The impressive seven storey building, in which concrete was used for the first time, boasts many celebrities and royalty among its guests, including the painter Claude Monet and the actor Charlie Chaplin.

Victoria Embankment and the South Bank

Strand, Somerset House & Courtauld Galleries

Somerset House overlooking the Thames

Somerset House

On the site of the original palace built 1547-50 for the Duke of Somerset, protector of England, the present riverside building was designed by Sir William Chambers and completed in 1786. It was erected to house government departments and learned societies and is still occupied by the Inland Revenue and the Registrar-General of Births, Deaths and Marriages. The superb south facing elevation overlooks the River Thames and has a terrace 800ft long resting on a series of 21 fine arches. Behind the north Palladian facade there is a huge quadrangle surrounded by similar buildings. The statue of George III was erected in the courtyard in 1788.

Courtauld Institute Exhibition

In 1990 new galleries were opened at Somerset House, once home to the Royal Academy, housing permanently the Courtauld Institute paintings. Famous the world over, this superb collection includes some of the best loved Impressionist paintings to be found anywhere.

The Strand

In Stuart times, the Strand was a riverside walk in the West End, bordered by mansions and gardens down to the River Thames. Their names still survive in the surrounding streets such as Bedford, Buckingham and Villiers Streets, and 5 Strand Lane is reputed to be a Roman Bath which was restored in the 17th century.

The Roman Bath

Situated in Strand Lane, this plunge bath is of red brick fed by a spring. It is thought to possibly be a reservoir restored by the Earls of Arundel for use in domestic offices of Arundel House.

York Watergate

This marks the position of the north bank of the River Thames before the construction of the Victoria Embankment in 1862. It was built by Nicholas Stone in 1626 as the watergate to York House. The arms and motto are those of the Villiers family.

Walk along Strand and Fleet Street

The historic route from Westminster to the City of London is along the Strand and Fleet Street. Places of interest are located along and just off these streets.. The places are to the right or left as you walk ahead. The walk starts at Trafalgar Square and ends at St. Paul's Cathedral.

(1) Trafalgar Square and St Martin-in-the-Fields Church.
(2) Charing Cross Station along the Strand.
(3) Villiers Street Shops and Watergate.
(4) Adelphi Theatre and Terrace, off Adam Street, where the actor Garrick and the writer Bernard Shaw lived.
(5) Savoy Hotel, Chapel and Theatre.
(6) Somerset House (1786) and Courtauld Art Galleries adjacent to Waterloo Bridge.
(7) St Mary-le-Strand (1714).
(8) Aldwych, BBC Bush House and India House, north side of the Strand.
(9) St Clement Danes Church (1680).
(10) Royal Courts of Justice.
(11) The Temple and Childs Bank (1671) along Fleet Street.
(12) Record Office and Museum with the Domesday Book of 1086, on north side of Fleet Street.
(13) Prince Henry's Room, timber framed building (1610) along Fleet Street.
(14) Temple Church, dates to the crusaders, Knights Templar in 1165 on south side of Fleet Street towards the Embankment.
(15) Inns of Court for legal professions.
(16) St Dunstan in the West (1671).
(17) Dr Johnson's House (1758) in Gough Square, off Fleet Street.
(18) Old Cheshire Cheese Public House.
(19) St Brides Church (1671) in St Brides Lane, south of Fleet Street.
(20) Past Ludgate Circus and along Old Bailey, the Central Criminal Court (1907).
(21) St Paul's Cathedral (1708). The crypt contains the tombs of Lord Nelson, Duke of Wellington and Christopher Wren.

Covent Garden and London Transport Museum

Covent Garden
This fashionable area of Central London has outstanding architectural features and is a pleasurable meeting place for both Londoners and visitors. Since the conversion of the old Market Hall to a complex of shops it has become a centre for arts and crafts shops, colourful restaurants, trendy wine bars and the London Transport Museum. At night, people flock to the nearby theatres and the famous home of opera and ballet - The Royal Opera House.

Covent Garden Old Market in WC2, was originally designed by Inigo Jones as a residential square in 1638. It was until recently the home of market stalls selling vegetables and flowers. It houses the Royal Opera House, publishers and the fine Church of St. Paul's. Some of the old market halls were built 1831-33 by Charles Fowler.

London's first square, this beautiful Italianate piazza was designed for Charles I and is bordered by superb 17th and 18th century buildings. It includes shops with covered walkways and the Church of St. Paul's, which has associations with artists and actors. The market place started in 1670 and grew until it was moved in 1974 to Nine Elms south of the river. It is now a precinct of shops, restaurants and offices. During the adaptation, new courtyards were excavated to open part of the basement for public use and to increase retail space. Covent Garden is one of the more famous tourists' attractions in London with the same number of visitors as the Tower of London. Thousands of people flock there daily for its arcade specialist shops, old pubs, restaurants, theatres and opera house.

The Royal Opera House
The original buildings of the Royal Opera House and the Theatre Royal were commissioned by Charles II in 1660 as part of Covent Garden. The present buildings originated in the 19th century.

Royal Opera Arcade
This is one of London's earliest arcades built by John Nash in 1816. It has beautiful Regency style bow-fronted shops, glassed domed vaults and elegant lamps.

St. Paul's Church
The original church was built by Inigo Jones (1631-8) on the west side of Covent Garden Market. It was rebuilt after a fire in 1798. There is a deep portico supported by square pillars and two columns in the middle.

National Theatre Museum
The Theatre Museum is located in an elegant Victorian building just a few minutes walk from Covent Garden underground station. The galleries are arranged on two floors with displays drawn from the world's best theatres collections. They celebrate 400 years of liver performance from Shakespeare to the present. The Museum is in the centre of theatreland and its postal address is 1E Tavistock Street, London WC2E 7PA (071 836 7891).

London Transport Museum
This museum has exciting exhibitions on the history and development of public transport. There are Royal railway coaches, vintage trains and buses. The exhibition has a selection of old posters and graphic works. You can see a superb collection of historic London horse and motor buses, trams, trolley buses, underground trains and models. You can even drive some of the vehicles yourself! There is a well stocked with books and unusual souvenirs. It is within walking distance of Covent Garden tube station.

Covent Garden and Royal Opera House

Piccadilly Circus, Mayfair and Regent Street

Piccadilly Circus

Piccadilly Circus and Eros

Piccadilly Circus is the centre of London's West End and the heart of tourist attractions. Its name is said to be derived from a 17th century tailor who sold lace collars called piccadils. The circus was designed during the early 19th century by Nash as part of a plan to join Prince Regent's Mansion, Carlton House, with Regent's Park. It is estimated that the circular underground station handles about 30 million passengers a year. The central statue of Eros represents the Angel of Christian Charity. It was built in 1892 by Sir Alfred Gilbert as a memorial to the seventh Earl of Shaftesbury. The octagonal bronze fountain on a stepped platform is surmounted by a winged figure of Eros in aluminium. Recently, Eros and his bow and arrow have been renovated at a cost of £30,000.

Piccadilly Tube Line and Station

This Line was opened in 1906 by David Lloyd George, then President of the Board of Trade, and ran from Finsbury Park to Hammersmith. It was the longest tube line in London and acquired the first railway escalator. 'Bumper' Harris, a man with a wooden leg, was engaged to travel up and down all day to prove its safety to passengers! It was extended northwards to Cockfosters and westwards to Uxbridge by

1932. In December 1977 an extension was opened to Heathrow Airport making the route 40.5 miles long.

Regent Street

Part of Nash's 1810 proposal to connect Regent's Park, Regent's Palace and Charlton House. The central curved portion between Piccadilly and Oxford Circus, known as the Quadrant, was devoted to fashion shops; the remainder of the street, Portland Street, was meant to be residential. The Quadrant has sweeps of colonnades of cast-iron columns.

St George's Church

Designed by John James and built 1720-24, the church has a large portico of six corinthian columns, projecting into Maddox Street. The American President, Theodore Roosevelt was married in this church in 1886.

Mayfair

This fashionable area of London lies between Hyde Park and Piccadilly. At Grosvenor Square stands the American Embassy with its golden eagle. Mayfair has many high class hotels such as Claridges, The Dorchester, The Hilton and The Ritz adjoining Green Park. It is also the home of the Royal Academy, Sotheby's and the Museum of Mankind.

Burlington House

Located in Piccadilly, this is the home of the Royal Academy and the Royal Society which was founded in 1660 as an independent learned society for the promotion of the natural sciences, including mathematics and all applied sciences of engineering and medicine. The 18th century palace was refaced during the 19th century.

Museum of Mankind

Located in Burlington Gardens, W1, the museum contains large collections portraying the cultures of Africa, The Americas, some parts of Asia and Europe, Australia and the Pacific Islands (071 437 2224).

Guinness World of Records

This museum in The Trocadero, Piccadilly Circus, has a permanent exhibition designed to bring to life by the use of models, videos and computers, facts and figures contained in the Guinness Book of Records. You can even stand next to the world's tallest man! (071 439 1791).

St James' Church

Wren's only church outside the City of London and built 1676-84 in the district of St James', the church has a large hall with galleries for additional seating.

Soho, Carnaby Street and Leicester Square

Eros, Piccadilly Circus

Soho and Leicester Square

A fashionable area of London, it has a high concentration of Italian and Chinese restaurants and delicatessens. Bounded by Regent Street, Oxford Street and Charing Cross Road, close by are the West End's theatreland and cinemas. During the summer the Soho Festival is very popular and there is heavy betting on the Waiters' Race. Many clubs and cabaret establishments are situated in this area which is also a centre for the fashion industry.

Carnaby Street

This narrow street in Soho has been world famous as the centre of young fashion and modern London since 1957. It is packed with colourful boutiques and hung with banners across the street. There are many excellent clothes shops with nearby restaurants and pubs.

The Odeon at Leicester Square

The building with dark granite facade was opened 1937, when there was considerable development and expansion of the media, mass circulation newspapers, radio broadcasting and cinema. American influence in the style of dress, music, dancing and films penetrated British life between the two World Wars. This had an extraordinary impact on a society which a mere two decades earlier had been Victorian in values and attitudes.

Walk from Trafalgar Square to Oxford Street

This walk starts from Trafalgar Square and ends at Oxford Circus, following St Martin's Place to Leicester Square then to Piccadilly Circus through Coventry Street and finally through Regents Street.

(1) National Gallery (1838), with its great art treasures.

(2) St Martin-in-the-Fields Church (1722),

(3) Along St Martins Place the National Portrait Gallery with its superb paintings.

(4) London Coliseum Theatre for English National Opera at the corner of William 1V Street.

(5) Leicester Square with cinemas(including the Odeon), and restaurants.

(6) The famous Hippodrome with disco.

(7) Turn left into Coventry Street towards Piccadilly Circus and Guiness World of Records is in the Trocadero.

(8) Piccadilly Circus and Eros.

(9) Walk northward into Regent Street with many shops including Jaeger (Woollens, etc), Scotch House (Cashmere, Shetland) Mappin & Webb (Silver and jewellery), Robinson & Cleaver (Linen), Garrard (Crown Jewellers), Austin Reed (men's clothing).

(10) Turn right into Beak Street leading to Carnaby Street, the fashion street for young persons clothing.

(11) Go into Great Marlborough Street and proceed to the corner of Regent Street where Liberty the fabric and fashion shop stands,

(12) Follow Regent Street to Oxford Circus and the world famous Oxford Street with its great department stores.

Carnaby Street

Oxford Street, Selfridge's, Foyle's and Smiths

Marble Arch and Oxford Street
Designed by John Nash in 1828, the arch was originally the main gateway to Buckingham Palace but was relocated to its present site at the west end of Oxford Street. It is near the old site of Tyburn Gallows!

Selfridge's
An Edwardian building located near Marble Arch opened by the American millionaire Gordon Selfridge in 1909 directly above the Central Tube Line. The beautifully decorated Ionic columns of the facade are set forward from the three intermediate floors by metal panels. This most impressive department store has everything the ladies need and is a great family shopping place. It has numerous departments with restaurants, hairdressing salons and other services. When first established, it was dedicated to women.

Marks & Spencer
In 1896 Mr Michael Marks entered partnership with Mr Thomas Spencer and opened up the "Penny Bazaars". In the 1920s they moved into the West End where a Marks & Spencer department store was established.

Whiteley's
This superstore was established in 1887 as a centre of arts and industry of the nation.

W.H. Smith
The bookshop firm was a multiple retailer directly associated with public transport with its chain of railway station stores, which later expanded all over London and the rest of the country.

Peter Robinson
Was opened as a linen drapers and expanded from one shop in 1833 to take over five adjacent shops by 1860.

Bon Marche
One of the first department stores, it was opened in Britain in 1877 and named after a store in Paris.

John Lewis
The John Lewis Partnership is one of the largest retailers in Britain. The retailing business started in 1864 with one shop owned by John Lewis specialising in the sale of textiles, which are still a most important part of their trade.

Centre Point
Built 1965-66, this spectacular 35-storey office block is of pre-cast concrete cladding and is a landmark at the eastern end of Oxford Street opposite Tottenham Court Road tube Station. It is one of the highest points in London commanding an excellent view of Central London.

Foyle's Bookshop
Situated in Charing Cross Road near Centre Point this largest London Bookshop was opened in 1906 by two brothers, William and Gilbert Foyle. It is still owned by the Foyle family and supplies an extensive range of educational books to overseas countries.

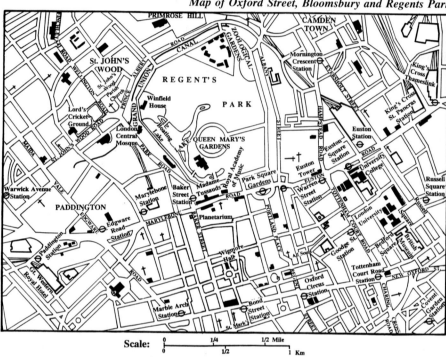

Map of Oxford Street, Bloomsbury and Regents Park

Scale:

Selfridge's, Oxford Street

British Museum and British Telecom Tower

Bloomsbury and Holborn

Built by the Victorian developer Thomas Cubitt in the middle of the 19th century, Bloomsbury consists of elegant Georgian houses and squares including Bedford, Russell and Tavistock Squares. This part of the West End extends from Euston to High Holborn and is the seat of scholarly activities including Gray's Inn, one of four Inns of Court of legal London. The British Museum and the University of London buildings are located within a short distance of Russell Square, Tottenham Court Road and Holborn tube stations. The British Museum currently houses millions of books of the British Library.

British Museum

This museum is one of the largest and finest in the world and its Exhibition Galleries are open daily, entrance is free. They contain unique and famous collections of Assyrian, Egyptian, Greek, Roman, British, Asian and Oriental antiquities. The Egyptian mummies are one of the outstanding collections in the museum. Others include the Assyrian bulls and lions in the Nimrod Gallery, the Benin, Cambodian and Chinese collections. The new mezzanine floor of Gallery 3 has marble sculptures which come from Eastern Greece and date back to the 6th Century B.C. The sculptures are placed on textured concrete plinths. Russell Square or Tottenham Court Road tube stations are within walking distance (071 323 8599).

The present Museum building was designed by Sir Robert Smirke, the quadrangle was completed in 1847 and the Reading Room by his son Sidney Smirke opened ten years later. The original museum was at Montague House on the same site opened in 1759 and contained the Sloane's Library and the Harley Collection. The Royal Library donated by George II and •the Cotton Collection were later added. The museum's vast treasures also include two of the four copies of the Magna Carta, the Elgin Marbles and the Rosetta Stone which led to the deciphering of Egyptian hieroglyphic writing.

The glass domed Reading Room and Library has seating for 400 people and has 25 miles of book shelves. The dome is magnificent, 42m diameter and 32m high. Karl Marx wrote Das Kapital at seat No. G7. It is statutory for publishers to send the library a free copy of any book printed in the UK.

University of London

The University of London Senate House was built 1936 in Bloomsbury to the north of the British Museum. The University consists of four separate colleges. Along Gower Street is University College with its classic architecture dating from 1828 by William Wilkins who also designed the National Gallery in Trafalgar Square. In the main hall you can see the clothed skeleton of philosopher Jeremy Bentham, one of the founders of the college. With the backing of the Archbishop of Canterbury, King's College was opened in the Strand. In 1836 the University of London was established as an examining and degree awarding authority for the London Colleges which lasted until 1900 when a federal system of colleges was established which included Imperial College, established at South Kensington in 1906, King's College and University College. Queen Mary College in East London joined at a later date.

British Telecom Tower

Built by the Ministry of Public Building and Works in the early 1960s, the 580 feet tower houses telecommunication equipment with revolving restaurant and a sightseeing gallery. The graceful structure in Howland Street, W1, has the most spectacular panorama of London. Unfortunately, due to bomb threat, the tower is currently closed to visitors.

Pollock's Toy Museum

Originally constructed between 1746 and 1769, this historic building of four stories is near the modern Telecom Tower. There are collections of toys, dolls and miniature theatres dating from Victorian times to the present day.

The Egyptian Gallery, British Museum

Madame Tussauds, Regents Park & London Zoo

Madam Tussuads and Planetarium

Regents Park and St. John's Wood

This Royal Park of 470 acres is one of the finest in Central London and contains the famous London Zoo. During the 16th century, it was part of Henry VIII's hunting forest. In 1811 the Prince Regent planned to connect the park and a new palace via Regent Street to Carlton House. The design of the park and its surrounding area was carried out by John Nash during 1812-26. It is surrounded by beautiful Regency terraces and imposing gateways. Through it passes Regents Canal and a boating lake with many species of birds including ducks, swans, herons and kingfishers. The Queen Mary's rose garden within Nash's inner circle provides a perfect place to relax. There is also an open-air theatre. At its northern end is London Zoo. On the west side is the elegant St John's Wood, well known as the home of Lord's Cricket Ground.

London Central Mosque

Situated on the edge of Regent's Park the Central Mosque, incorporating a steel dome with copper finish, is just one of many buildings of international standing. Its golden dome is the largest in the UK.

London Zoo

One of the largest and most famous zoos in the world containing a huge collection of animals including the giant panda, gorillas, sea lions, elephants, camels, etc. Feeding time for the lions and tigers is an exciting spectacle. The zoo was laid out by Decimus Burton in 1827 and has been modified over the years by a number of architects involved in the design of animal houses. The aviary was designed by Lord Snowdon.

The Aquarium

Both well lit and professionally displayed the aquarium is full of sea and river fish and amphibians from Africa, Asia, Europe, etc. There are cobras, octopus, stingrays and sharks.

Apes and Monkeys Pavilions

There are five pavilions each containing service areas and dens from which the animals have direct access to the enclosures.

Marylebone

South of Regents Park is Marylebone with its attractive buildings and imposing Victorian squares. Marylebone High Street with its trendy shops and food stores is the main route through the area. Baker Street traverses the area north to south and is associated with the fictitious detective, Sherlock Holmes and his assistant Dr Watson who were supposed to have lodged at No. 221B, now owned by the Halifax Building Society. Running parallel to Baker Street is Harley Street with private medical consulting rooms. Nearby is Wigmore Hall where many musicians have made their London debut. Broadcasting House, the headquarters of the BBC stands at the southern end of Portland Place.

Heinz Gallery

This is part of the Royal Institute of British Architects. The design sponsored by Mr & Mrs J Heinz and built in 1972.

President Kennedy Bust

At the eastern end of Marylebone Road, there is a bust of the famous American President John F Kennedy (1927-1963).

Madame Tussauds Waxworks

The exhibition shows many past and present personalities, including royalty, statesmen, politicians, sport and entertainment stars. Madame Tussaud's self-portrait stands at the entrance. The Chamber of Horrors features notorious criminals and executions and a tableau of the Battle of Trafalgar. Louis XVI and Marie Antoinette are also shown. Madame Tussaud showed her waxworks first in Paris in 1770, then came to London in 1802 and opened her museum in 1835. The latest addition to the museum is the "Spirit of London", a five minute ride through 10 scenes - featuring lifesize speaking figures - depicting the City's history from Tudor times to the present.

The Planetarium

The Zeiss Star Projector takes you through time and space in daily shows. In the evening laser light concerts are held to the best of rock and pop music.

Lord's Cricket Ground

The ground of the MCC is located in St John's Wood Road, NW8. The Mound Stand provides 4,400 seats with bars, shops and public facilities facing onto a colonnade. Above the mound, a steel structure supports two upper tiers.

All Saints Church

In Margaret Street, off Marylebone, this small Gothic style church was built by W Butterfield (1849-59) of dark red brick with stone enrichments. The Victorian interior has ornaments, marble, coloured tiles and stained glass.

The Law Courts, Grays Inn and Lincoln's Inn

Legal London
The Strand and Fleet Street stretch from Charing Cross to Ludgate Circus and are associated with London's legal professions, but previously it was also the centre of the newspaper and printing industry which have now been relocated in London Docklands. On both sides of the Strand are the four Inns of Courts - the Middle and Inner Temple, Lincoln's Inn and Gray's Inn. The Royal Law Courts front the Strand and unite the four Inns.

The Law Courts
In 1862, the Inns of Court were centred around Chancery Lane. There were over 1,500 resident Barristers, 3,500 Solicitors and more than 5,000 law clerks and court officers employed in London's legal district in 1851. The superior courts for civil law were brought under one roof with the opening of the Royal Courts of Justice (the Law Courts) in The Strand in 1882. Designed by G E Street, the imposing stone faced building has over 1,000 rooms. The building, designed in the Gothic style, was opened in 1882 and is the second highest court in the land after the House of Lords. The 34 courtrooms are arranged in a number of divisions: Courts of Appeal, Queen's Bench for Common Law cases, Chancery Courts for Equity Cases, and Family Courts. The Central hall with a fine mosaic floor, is open to the public who also have access to the galleries from where they can follow the proceedings in the court rooms.

St Mary-le-Strand
Opposite the Law Courts, this Baroque Church by J Gibbs (1714-17) is located in the middle of an island in the Strand. On the west side there is a semicircular entrance porch. Above the central window flanked by two columns there rises the steeple with three stages of columns and pilasters.

Dickens' House
This historic house is a museum with a library and gallery devoted to information and research on the life and books of Charles Dickens during the first half of the 19th century. The museum is at 48 Doughty Street, WC1 (071 405 2127). The nearest tube station is Holborn.

Grays Inn
It has been an Inn of Court since the 14th century and is approached from a passage next to 22 High Holborn. The 16th century Hall and other buildings were restored following the bomb damage of World War II.

Lincoln's Inn
In the centre of London, Lincoln's Inn is one of the four Inns of Court which are unincorporated bodies of lawyers who have the power to call to the Bar those of their members with a rank or degree of Barrister-at-Law. Its history dates back to 1422. The other Inns are Middle Temple 1501, Inner Temple 1505 and Gray's Inn 1569. The Old Hall is one of the finest in London. It was here that Sir Thomas More spent much of his professional life after joining the Inn in 1496. In the past, the Hall was used as a court of justice. Throughout the legal year the adjoining Chapel, completed in 1623, holds services on Sunday mornings and the general public are welcome to attend. The gatehouse from Chancery Lane was built 1521.

Sir John Soane's Museum
Built by John Soane for his own occupation (1812-14) with the front facing Lincoln's Inn Fields, it contains his private collections of paintings, drawings and sculpture.

Lincoln's Inn Fields Netball
The spacious square with lawns and trees is bounded by Lincoln's Inn and rows of 18th and 19th century houses. Excellent ladies' netball teams compete in the Fields during the lunch break of the City workers. Holborn tube station is nearby.

Sadler's Wells Theatre
This is the original home of the Sadlers Wells Ballet Company and Sadlers Wells Opera Company. The ballet company now performs at Covent Garden Opera House.

Prudential Assurance Company
The imposing red brick headquarters of the Prudential Assurance Company in Holborn was designed by Alfred Waterhouse and erected 1879.

Drury Lane Theatre
Built by B Wyatt (1810-12), it is one of the best 19th century theatre buildings in London. Apart from the addition of a portico and Ionic colonnade at the side in 1820 and 1831 respectively, the beautiful original interior and exterior are intact.

The Old Curiosity Shop
Built in 1567 and made famous by Charles Dickens in his book bearing the same name, this old world building is in Portsmouth Street, close to Lincoln's Inn Fields. It is used as an antique and modern art shop.

The Law Courts

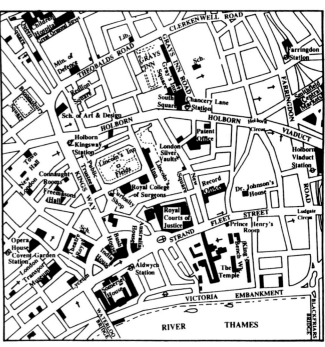
Map of Legal London

THE CITY OF LONDON

The City of London

The "Square Mile" of the City is the oldest part of London, dating back over a period of 2000 years. It is built on two hills, Ludgate Hill and Cornhill. Remains of the Roman Wall which once surrounded it are still evident in many locations. You can still see traces of the medieval street plan with narrow alleyways and old churches which survived the Great Fire of 1666. These are surrounded and dwarfed by the modern skyscrapers and new office blocks. The waterfront has been redeveloped since the removal of the docks to Tilbury.

All year round tourists flock to the Tower of London with its Crown Jewels and uniformed Beefeaters. In the early mornings, the historic meat market at Smithfield is a centre of activity. A little later the railway stations, streets and office buildings are full of City gents and other office workers engaged in international banking, insurance and stockbroking. After the bustling daily activities, the City becomes calm and peaceful in the early evening except for the area around the Barbican Centre with its famous art and music centre.

Fleet Street, Temple and Prince Henry's Room

Fleet Street

From medieval times Fleet Street has been the main road approaching the City of London from Temple Bar to Ludgate Circus and has been the home of journalists and newspapers for many centuries. The old Fleet River is sadly now a sewer beneath New Bridge Street and discharges into the Thames at Blackfriars Bridge. Until recently, Fleet Street was the home of the newspaper industry. The "Street of Ink" had been associated with printing since the 15th century days of Caxton. Up to 1984 all national and most provincial newspapers had their offices in or near it. Since then, practically all the London and national papers have been relocated in various parts of London Docklands at Wapping, Isle of Dogs and Surrey Docks.

Prince Henry's Room

Built as a tavern in 1610-11, this building in Fleet Street was later altered and given a false frontage. It was acquired by the London County Council in 1900 and restored to its original façade. At ground level it has a modern shop front and rusticated stone archway leading to the Inner Temple bearing the date 1748. The interior contains a late 17th century stair and an early 17th century room called Prince Henry's Room on the first floor over the gateway. The enriched plaster ceiling of the room incorporates the emblem of the Prince of Wales against the background of oak panelled walls.

Middle Temple Hall

The hall, built 1562-70, has the finest double hammer beam roof in the country and has a magnificent screen at its lower end comprising five beautifully carved sections. At the western end hang portraits of Queen Elizabeth, Charles I, Charles II, James II and others up to George I. Sir Francis Drake dined here on 4 August 1586 after arriving back from his exploration round the globe. Permission is required to visit the hall. The Middle Temple's library, established 1641, is an excellent reference on American and European laws.

Temple Bar

Fleet Street, EC4. This marks the western boundary of the City of London. It is where on Royal visits to the City, the Lord Mayor meets the Sovereign and surrenders the City's pearl sword.

Temple Church

The round church has a circular nave built 1160-1185 in the classical style of the knights templar, and a chapel added 1220-40; both were restored after the Second World War.

Blackfriars Bridge

Built 1860-69 by J Cubitt and H Carr, this iron bridge has five arches resting on granite piers. Queen Victoria opened it on the same day as the Holborn Viaduct. The west side of the bridge was extended in 1910.

Public Records Office

Located in Chancery Lane, WC2, this building is the storehouse of the State's documents and records preserved in fire-proof chambers. The public search-rooms and museum are open daily. The museum contains many valuable and interesting documents such as the original Domesday Book and many autographs and letters.

Dr Johnson's House

The 17th Century house in Gough Street off Fleet Street, was the home of Dr. Johnson from 1748-59 and contains relics and contemporary portraits (071 435 2062). It is close to Temple tube station.

Cheshire Cheese

Built in the late 18th century of an oak construction this famous public house was frequented by Galsworthy, Charles Dickens, Dr Johnson, Boswell and Garrick. A parrot named Polly was once a well known resident who on Armistice Night 1918 imitated the popping of a champagne cork 400 times and fainted. When he died at the age of 40 his obituary appeared in many newspapers. The inn is near Blackfriars tube station.

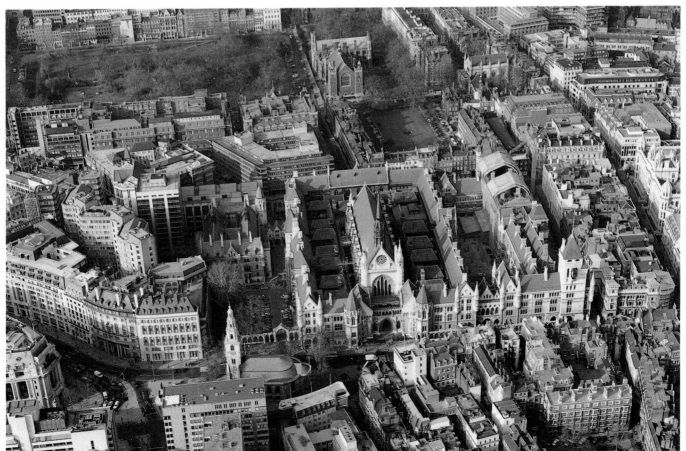

Lincoln's Inn, Law Courts and the Temple

St. Paul's Cathedral - Wren's Masterpiece

St. Paul's Cathedral

The original Cathedral was founded in 604 and rebuilt in stone by the Saxons later in the same century. Further rebuilding took place during the 10th and 11th centuries when fire damaged part of the City. In the 12th century the land around the Cathedral was bought by the Bishop of London and some donated by Henry I. The area was enclosed by a wall. The precinct contained the Chapter House, the Cloister, the Bishop's Palace, the Charnel House (the mortuary) and Paul's Cross - the City's principal public meeting place. There was an open-air pulpit from which sermons and proclamations were made; this was demolished by Parliament in 1643. The Medieval map of 1377-99, during the reign of Richard II, shows clearly the precinct of St Paul's and the pattern of roads and alleyways.

Christopher Wren's Masterpiece

After the Great Fire of London in 1666 Sir Christopher Wren undertook its complete re-building between 1675-1710. The Gothic Cathedral was designed around a central lantern about 364 ft high, with elegant twin towers and an extensive vault. The front elevation has a central two-storeyed portico, flanked by two fine Baroque towers. The splendid hemispherical dome is supported by a colonnade all round the circumference and is decorated by frescoes based on the life of St Paul by James Thornhill (1716-19). The fine choir stalls are the masterpiece of Gibbons. The chancel, with aisles of three bays, is the same length as the nave. It comprises many externally sculptured details, and wrought iron screens. It is based on St. Peters in Rome but on a significantly smaller scale.

Queen Anne's Statue

Erected in 1886 in front of St. Paul's Cathedral, this is a replica of the original by Bird, 1712. Surrounding the Queen are figures which represent England, France, Ireland and North America.

Amen Court

Original wallpaper dating from the 17th century was found when refurbishment work was being carried out to the buildings behind St. Paul's Cathedral in 1992. Hand-painted wallpaper was also found in the downstairs room of 1 Amen Court, occupied by the Dean and Chapter. The building is believed to be the oldest timber framed building in London. Built in 1694 the front is of red brick with square headed windows.

Charterhouse

A large 16th century mansion house built on the site of the Carthusian Priory originally of the 14th century. The fine Elizabethan Great Hall, circa 1571, has a beautiful fireplace and ceiling.

College of Arms

Established by Edward I in the 13th century, it is the Home of Royal Heralds. The building, which was restored during the 17th century, contains charters, rolls and records. The College examines pedigrees, and designs and grants new Coats of Arms. It is located close to St Paul's Cathedral.

Museum of Telecommunications

Located at 135 Queen Victoria Street, EC4, this is Britain's Telecom Technology Museum and showcase, which is worth a visit. It is suitable for educational visits.

Mermaid Theatre

The first new theatre to be built in the City for 300 years it was opened by the Lord Mayor on 28th May 1959. A conversion of a Victorian riverside warehouse at Blackfriars, Puddle Dock in Upper Thames Street, EC4.

St. Pauls Cathedral

Old Bailey, Paternoster Square and Smithfields

Paternoster Square

The area north of St Paul's Cathedral, known as Paternoster Square was developed in the 1960's following war damage with modern buildings including Courtenay House and Sheldon House. A new plan has been prepared by Paternoster Associates for the redevelopment of the area with traditional buildings. The proposals include six new office blocks over 80 shops of varying sizes and a public square. There will be glazed arcades and streets with shops leading into and surrounding the edges of the square with formal pavilions. The Chapter House will be restored to its original context, part of the streetscape of the realigned St Paul's Churchyard, flanked by buildings planned on the pre-war configuration. The plan includes new links to St Paul's underground station.

The Old Bailey

Situated in Old Bailey Street, this impressive Central Criminal Court was opened in 1907 to the design of Mountford. It incorporates a 60m high central dome surmounted by the statue of Justice. Most of the notorious criminals of the 20th century have been prosecuted here. The facade was built from stone following the demolition of Newgate Jail which existed on the site until 1901 and was the place of many public executions since 1783. Each year in January, the Lord Mayor of London leads a procession from his official residence, Mansion House, to the court to open a new session. There is a public gallery where visitors can watch the judicial proceedings.

St Bartholomew The Great

Founded in 1123 as an Augustinian Priory with an adjoining hospital, the Norman chancel has circular piers with moulded capitals and a gallery of four arched openings. The architecture is similar to that

Interior of Old Bailey

of the Chapel of St John at the Tower of London.

Smithfield Market

The ten acre meat market is the largest in the world. The buildings were erected between 1868 and 1899. The district is an historical site of public medieval executions, a cattle market and the Bartholomew Fair. In the north east corner is the original Tudor gate house built over a 13th century archway leading to the Church of Bartholomew the Great. The south east side is occupied by St. Bartholomew's Hospital, the oldest hospital in London, being founded in 1123. The Gateway of 1702 bears London's only statue of Henry VIII. The traditional city market has arcaded avenues of raw meat and four large halls organised around the Grand Avenue.

National Postal Museum

King Edward Street EC1. Founded in 1965 to commemorate the history of British Postage Stamps of the reign of Queen Victoria and of creating a national home for postal history. The museum contains the Phillips Collection of 19th century British postal stamps and the registration sheets for almost every postage label issued since the commencement of penny postage in 1840. The recently acquired stamp collection of rock star Freddie Mercury, is also on display (071 239 5420).

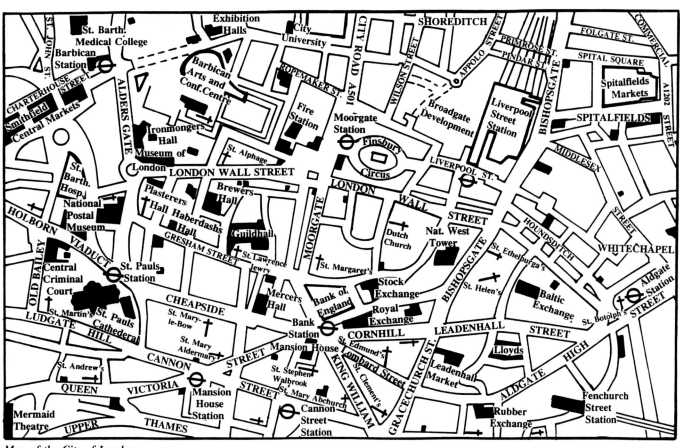

Map of the City of London

Museum of London and the Old Roman Wall

Museum of London

The Museum of London is considered Britain's top social history museum. Its origins can be traced back to two older museums; the London Museum founded in 1912 and housed at Kensington Palace, and the Guildhall Museum founded in 1826 by the City of London Corporation. The two museums were amalgamated in 1965 and moved to a new purpose built property at London Wall, opened by Her Majesty the Queen on 2nd December 1976.

The permanent exhibition and live displays show the various themes recurrent throughout London's history and the everyday lives of its people such as housing, dress, trade and commerce, religion, art and leisure. The story is told chronologically covering prehistoric times, Roman, Medieval, Tudor, Stuart, Georgian, Victorian and 20th century periods. The photographic department has an extensive record of change of contemporary London. New galleries are constantly created for important collections of paintings, pictures, costumes, etc. not only to delight the visitors but also to provide an educational service to schools and colleges.

The permanent collections represent 2000 years of London's history ranging from Roman times to the 20th century. The collections are wide ranging and include paintings, photographs and prints, excavated objects, items of industrial artifacts and costumes. The galleries are designed for maximum flexibility allowing you to start at the beginning or in any other gallery depending on the period of history that may be of particular interest to you.

London Wall

London Wall is the street running from Aldersgate Street to Broad Street following the line of the old city wall which the Romans built in the late second century. Nearly two miles long and 18ft high, the wall enclosed 330 acres of the city with a defensive ditch outside. There were six main gates: Aldgate, Aldersgate, Bishopsgate, Cripplegate, Ludgate and Newgate with smaller pedestrian entrances at Aldermanbury and the Tower of London along the River Thames. Excavations in 1977 revealed traces of the Thameside defences.

Roman Wall Preservation

The wall was built of Kentish ragstone brought by river from the Medway. In 1962 a barge was discovered at Blackfriars Bridge with a cargo of ragstone believed to be for the construction of the wall. The barge was a flat bottomed 55 foot long craft. A coin in the barge's mast stepping hole (still a common launching practice today), suggests it was about 100 years old when it sunk.

At a later date more than 20 defensive bastions were added and the wall was maintained by the Romans until their retreat in 410 AD. The Saxons and Normans improved it for about eleven centuries. During the 16th century the City of London expanded into the surrounding countryside and the ditch was neglected and became a rubbish dump. Architects and builders later incorporated the remains of the Roman wall into the foundations of the Regency and Victorian buildings which still exist today. Sections of the wall are preserved near the Museum of London at the Barbican complex. Other surviving parts of the wall are preserved along St. Alphage on the north side of London Wall, St. Giles Churchyard, Cripplegate, Jewry Street, off Trinity Square and in the Tower of London. (see page 9).

Little Britain

A new attractive building in a series of terraces facing the Museum of London.

Museum of London

Barbican Centre, Wesley's Chapel & Broadgate

The Barbican Centre

Barbican Art Centre

The Capital's well known Arts and Conference centre was completed in 1982 and is situated conveniently for the Barbican and Moorgate tube stations. The complex is in reinforced concrete with prefabricated components. The pond and cascading water falls are attractive features. One of Europe's largest arts centres, the Barbican was named after a medieval stronghold and developed by the Corporation of London following war damage. It is a great art centre and includes a concert hall, the home of the London Symphony Orchestra. The Arts Theatre is also the home of the Royal Shakespeare Company. There are also the New Guild Hall School of Music and Drama, the City of London School for Girls, the Museum of London and a section of the old Roman Wall. The three residential 44-storey towers are named after Cromwell, Shakespeare and Archbishop Lauderdale. There are three cinemas, a public library, conference rooms and the Barbican Art Gallery with its bookshop. In the middle of the complex is the historic church of St Giles Cripplegate where Oliver Cromwell was married in 1620.

Gresham College

Frobisher Crescent, Barbican Centre, EC2 (071 638 0353). This independent college arranges specialist public lectures on aspects of arts and science throughout the year, for lunchtime and evening sessions.

Wesley's Chapel

49 City Road, EC1Y 1AU (071 252 2262) John Wesley (1703-1791), the Father of Methodism, built his chapel and house in 1778 to the designs of George Dance the Younger. There are seven historic buildings, two courtyards, two chapels, two museums, two pulpits used by Wesley, two organs and John Wesley's tomb and statue. It is the Centre of Methodist's Church. The nearest tube stations are Old Street and Moorgate.

Broadgate Complex

Surrounding Liverpool Street Station this large complex of office blocks and squares was completed in 1991. Built using fast track techniques the offices were speculative developments in the early 1980s when demand was high. The Broadgate Arena hosts concerts during the summer months and converts into a skating rink during the winter. There are various works of art in the complex including a mural and a fine steel sculpture. Exchange House to the east of the site is a beautifully structured building with the arched façade spanning the railway tracks of Liverpool Street Station. The third public space, Exchange Square is similarly constructed.

Finsbury Avenue

A city office building of eight floors with a central top-lit atrium is located near Liverpool Street Station. The bronze colour external wall incorporates sun screens and a series of landscaped terraces. The courtyard has a sculpture by the American artist, George Segal.

Aldersgate Street Building

With available building land severely restricted, progressively deeper basements became a feature of London buildings in the City from the early 1970s. At 24m deep, Aldersgate Street building located adjacent to the Barbican complex has one of the deepest basements constructed to date.

City of London Guildhall and Mansion House

Medieval Guildhall

The 15th century building has a façade by George Dance, 1789 and later additions by Sir Giles Gilbert Scott. It is the City of London's Hall where the Lord Mayor is elected with his Sheriffs. The Great Hall is used for ceremonial occasions. Early in November each year Lord Mayor's Day is celebrated in London with a procession and banquet, when the Prime Minister of the day makes an important speech. The Hall was badly damaged during the air raids of 1940, when the fine roof and the famous figures of Gog and Magog were destroyed. New wood effigies of them were unveiled in 1953. The two legendary statues represent the sole survivors of a race of giants, offspring of demons and the daughters of Diocletion, who were brought as prisoners to London. They have often been paraded in processions and the Lord Mayor's show during the last five centuries. The crypts have medieval grained vaulting. There is an excellent Art Gallery and Library containing an unrivalled collection of materials on the history of London.

Guildhall Extension and Precinct

The municipal offices were built 1955-58 with light bricks and 1920s motifs along the ground floor. Completed in 1974, the L-shaped extension is clad in pre-cast concrete and similarly detailed with motifs of pointed heads and concrete umbrellas. The Precinct is connected to the Barbican complex at first floor level. Moorgate Station is nearby.

East Wing and Roman Amphitheatre

The discovery of a Roman amphitheatre beneath the site of the extension to the City of London Guildhall forced a major structural redesign. A massive 600ft steel transfer structure forms the design for the new East Wing extension on which work started in 1993. Heavy steel beams 1.5m deep transfer loads to eight piles positioned around the remains of the Roman amphitheatre. Design of this structure evolved after archaeologists discovered the remains of the amphitheatre in 1987. The amphitheatre resembles a similar one built by the Romans in Chester. The eastern extension will contain an Art Gallery and a Reception Gallery for the City of London (see page 49).

Clockmakers' Company Collection

Founded in 1813, the Clockmakers' Collection contains some of the works of the 18th century clockmaker John Harrison. He won a prize of £20,000 from the British Government for designing a perfect marine timekeeper in 1714. Other clocks and watches include Earnshaw's chronometer watch of 1791 which was in use at the discovery of Vancouver, Canada. The collection is housed in the Guildhall Library, Aldermanbury, E.C.2.

Mansion House

Situated near to Bank tube station this building is the official residence of the Lord Mayor during his year of office. Facing the Bank of England, the Palladian building was designed by George Dance, 1739. The new Docklands Light Railway extension to Bank passes beneath the building. The 18th century decor of the staterooms is superb. The Egyptian Hall, 27 metres long, accommodates 320 people and is the main banqueting room. The Lord Mayor's private apartment is at the top of the building. During the 15th century Richard (Dick) Whittington served three times as Mayor.

800th Anniversary of the Lord Mayor

In 1989, the office of Lord Mayor of London celebrated its 800th anniversary. In 1189, Henry Fitz Eylwyn was chosen by the burghers of London to be their representative and to preside over the City of London. His appointment founded the line of mayors that has survived for over eight centuries. The title was brought by the Normans who came to London during the 11th century. The election of the Mayor is by the 'Common Hall' and takes place annually in the Guildhall on Michaelmas Day, 29th September. People can see the Lord Mayor Elect as he comes out of the Great Hall, into the Guildhall yard. The Lord Mayor's show and colourful procession through the City takes place annually early November.

Old Dr Butler's Head

Nearby Guildhall, this inn was founded in the 17th century by Doctor Butler, a physician to the Court of James I. It was rebuilt after the Great Fire of 1666 and again after the Blitz of World War II.

The Guildhall

City of London Livery Companies

Livery Companies Halls

In the City of London there are one hundred Livery Companies, some of which are many centuries old. They began as trade guilds and associations of persons to regulate the various trades and protect their members and interests. They became very powerful in the government of the City and were recognised as suitable organisations to which benefactors would entrust charitable funds. They are called Livery Companies because, in the early 14th century, many of them assumed distinctive dress. The Butchers were known to have a hall outside the City wall as early as the 10th century but the oldest charter of 1155 is held by the Weavers' Company.

The associations, with members engaged in luxury goods and overseas trades, had greater standing and influence than others, and the present order of precedence was largely laid down in 1516 for the 48 companies in existence then. Eight of today's Great Twelve Companies provided 220 out of the 235 Aldermen of the City. Until 1742 the Lord Mayors were also required to be members of the Great Twelve. New Companies have been formed over the years and some amalgamations have taken place. They all have prestigious buildings within the City boundaries.

The degree of involvement of companies with their respective trades or industries today is variable. Some companies only act as administrators of charitable trusts and take part in the pageantry of the City. While others such as the Goldsmiths, Fishmongers, Vintners, Glovers and Stationers Companies have considerable involvement with their respective trades. Most of the Companies' Halls or offices are in historical buildings and are open to visitors.

The Great Twelve Companies

The Great Twelve Livery Companies with their addresses are:

(1) Mercers' Company, Ironmonger Lane, EC2
(2) Grocers' Company, Princes Street, EC2
(3) Drapers' Company, Throgmorton Street. EC2
(4) Fishmongers' Company, Foster Lane, EC2
(5) Goldsmiths' Company, Foster Lane, EC2
(6) Merchant Taylors' Company, 30 Threadneedle Street, EC2
(7) Skinners' Company, 8 Dowgate Hill, EC4
(8) Haberdashers' Company, Staining Lane, EC2
(9) Salters' Company, Fore Street, EC2
(10) Ironmongers' Company, The Barbican, EC2. The hall is situated adjacent to the Museum of London.
(11) Vintners' Company, Upper Thames Street, EC4
(12) Clothworkers' Company, Mincing Lane, EC3

City Information Centre

For further information on the Corporation, the City and the Livery Companies Halls, visitors should contact The City of London Information Centre, St Paul's Cathedral Churchyard, EC4M 8BX (071 606 3030).

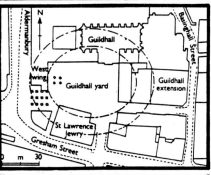

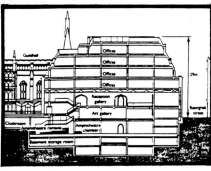

From top to bottom: (i) View of the construction of the Guildhall eastern extension, (ii) plan of the Guildhall with the Roman Amphitheatre superimposed, and (iii) a cross section of the extension

City of London Badge

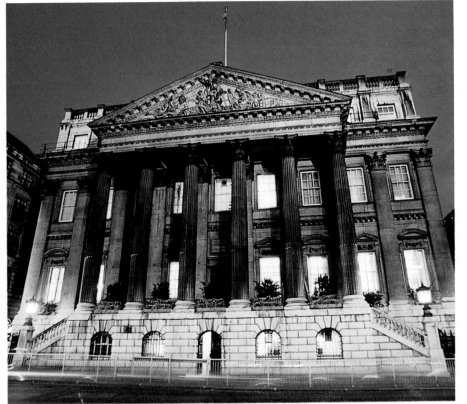

Mansion House at night

49

Bank of England, Museum and Royal Exchange

Bank of England

The Bank of England in Threadneedle Street, EC2, was originally built to the design of Sir John Soane, the architect to the bank from 1788 to 1833. The outer walls are still the originals but the interior was rebuilt by Sir Herbert Baker in 1937. The Bank is the centre of the United Kingdom monetary system and its activities are concerned with the carrying out of national policies and international banking transactions. It deals with income from taxation and is the sole printer of banknotes in England. New notes are issued either in return for old, or when backed by a new issue of securities. It holds the national gold reserves. The Bank stands at one of the busiest parts of London in the heart of the city. The vaults hold millions of pounds in gold bullion.

Bank of England Museum

The museum is next door to the Bank of England in Threadneedle Street, but the entrance is in Bartholomew Lane. This is where the Bank was founded and did not move to its more famous building until 40 years later. The Soanes' Bank Stock Office is in the style of a church interior and is a replica of how it looked in 1793. There is the Royal Charter granted by William and Mary in 1694 plus the finest collection of bank notes in the world. Around its perimeter stand showcases displaying the important phases in the Bank's history. In the centre of the Rotunda you are attracted to a showcase in which glints two gold bars!

Royal Exchange

Prior to the middle of the 16th century, the merchants of London met at St. Paul's. Sir Thomas Gresham, a powerful merchant, was impressed with the exchange in Antwerp, and built a similar one in London. Queen Elizabeth visited the building in 1568 and declared it the Royal Exchange. The merchants occupied ground level and the retailers the second floor. The building was destroyed during the Great Fire of 1666 and another one built which was also destroyed by fire in 1838. The present building was designed by William Tate and opened by Queen Victoria in 1844. The thriving Port of London trade kept the building busy for over a century. Following the closure of the London Docks, it has been used for insurance transactions. The grasshopper on the clock tower is the Gresham family crest.

In the historic Exchange, opposite the Bank of England, deals in commodities and insurance have been performed for centuries. In September 1981 a new market known as the London International Financial Futures Exchange was established in the building. Members include the major international banks, merchant banks and discount houses whose objective was to protect themselves against financial risks. The floor of the exchange offers a colourful spectacle as the role of the participants is indicated by the different coloured jackets worn.

London Stock Exchange

Old Broad Street, EC2. Founded in 1801, it is the market place for stocks and shares. A new tower block houses Stock-brokers' offices and the trading floor was opened in 1972. Its function is to provide an efficient and free market for the purchase and sale of securities and by so doing provides new capital for industry and commerce. Some 9,000 securities are listed, about nine times the Wall Street listings of New York. The business of the Exchange can be observed from the Visitors' Gallery.

The Stock Exchange is part of the life of many people in the United Kingdom and Overseas. More than ten million individuals own shares directly, while those who belong to a pension scheme or have an insurance policy are indirectly involved in the market.

82 Lombard Street

This banking headquarters is located on the apex of the triangular site at the junction of Cornhill and Lombard Street overlooking the Bank of England, Mansion House and the Royal Exchange. The whole site is a conservation area. In 1803 the buildings consisted of a group of small houses, one of which became the office of the Globe Insurance Company. Records show that in the past this busy corner had a lottery office and the ground floor unit in Cornhill was the bookshop of Thomas Guy, the founder of Guy's Hospital. The company expanded and in about 1835 moved.

Barclays Bank Headquarters

Rising 270 feet above Lombard Street is the new Barclays Bank's headquarters building in the centre of the City. Construction of the triple towered building started in December 1990 and was completed in April 1993.

Simpson's Tavern

This interesting tavern is located in an enclosure of courts and alleyways off Cornhill.

Bank of England and the Royal Exchange

National Westminster Tower & Baltic Exchange

London in the 1980s

During the 1980s, London's business and financial services burgeoned. It saw the rise of the service sector and the decline of manufacturing. As a leading world financial centre, the City of London was transformed by a building boom. The old stately buildings of The Bank of England, the Royal Exchange and others have been dwarfed by the modern architecture of the National Westminster Tower, Lloyds and other tall buildings.

National Westminster Tower

The new National Westminster Bank Head Office tower is 183m high and is shaped like a clover leaf in plan similar to the logo used by the Bank. The structure was designed to overcome complicated erection problems and one where steel was the only suitable material which could be used to give an economical solution. When completed in 1981 the Tower was the tallest block in Britain but is now second to Canary Wharf Tower in London Docklands. It is said that the design creates a fluid expression in the facades. Bank Station is close by.

Baltic Exchange

The Exchange is the only international exchange in the world where ships are found for cargoes and cargoes for ships. There are 600 associated UK registered companies who elect approximately 2,000 members to represent them on the floor of the Exchange. There they engage in the international trading of goods between nations by air or ship, together with the ownership and operation of ships and aircraft. The Exchange also houses fine futures markets, dealing with commodities such as grain, soya meal, meat and potatoes. Their main activity covers the provision of shipping for bulk cargoes such as grain, ores, coal, cement, fertilisers and scrap. In 1992 the building was sadly damaged by a terrorist' bomb.

Bush Lane House

The first building in the UK to use a water filled tube structure as the means of fire protection. The eight floor office block has an unusual appearance, set 9m clear of the ground, to allow for the subsequent construction of the London Underground Fleet Line below the site. The full height lattice structure of exposed stainless steel tubes was adopted for structural reasons to provide the maximum unimpeded floor area. Chemicals in solution have been added to the water in the tubes to prevent freezing. In the event of serious fire, the local rises in the temperature of exposed steelwork can be controlled by the circulation of water to compensate for losses due to evaporation.

National Westminster Tower

Lloyd's of London and Leadenhall Market

Lloyd's of London

Lloyd's Building

This impressive and unconventional type of office block is situated near Bank Station in the City. The lifts, stairs and services are placed in six satellite towers around the perimeter of the building. External cladding consists of triple glass through which warm air passes to improve thermal insulation. The pedimented entrance to the 1925 Lloyd's building has been retained. A large market place, called The Room, is a main attraction above which there is a central atrium. The building houses the Institute of London Underwriters, where insurance brokers have quick access from the ground floor via escalators and a spiral staircase to the underwriting rooms on the first, second and third floors. On the ground floor is the underwriting room with the casualty book, rostrum and Lutine Bell.

Lloyd's of London was established in the 17th century in a city coffee house by Edward Lloyd and today is a leading financial institution with underwriting capacity in excess of £11 billion. More than 20,000 underwriters are associated with the firm, grouped into 430 syndicates, with nearly half involved in aviation and maritime risks. The whole organisation provides the world's greatest insurance market, involved in major projects such as North Sea oil platforms and the Channel Tunnel.

Lloyd's of London Exhibition

Visitors can ride a lift up to a multi-media exhibition, tracing the history of Lloyds and insurance business to the present day. There is also a viewing gallery overlooking the Underwriting Room and you can gaze up into the impressive 200 foot atrium.

Commercial Union and P&O

These two modern buildings were built 1964-69. The Commercial Union building near Bank Station is a contemporary office block with immaculate detailing. The building is 387 feet high and 124 feet square in plan. The design was the outcome of a brief requiring maximum clear floor span and a minimum number of columns through the ground and basement floors. The P&O complex has a tower and podium with shopping units at ground level.

Leadenhall Market

In Gracechurch Street, this Victorian glass and iron structure hall of 1881 houses the poultry market. Designed by Sir Horace Jones, the retail market contains some seventy shops devoted to the food trade. Beneath the market lies the site of a Roman Basilica built during the 1st century following the landing of the Romans in London in 43AD.

London Metal Exchange

The Metal Exchange is the most important market for commodity prices in the world. The members meet twice daily just off Leadenhall Market to transact business in metals and many of these transactions are from overseas orders.

Commodity Exchange

The London Commodity Exchange, a merger of the former Commercial Sale Rooms and the Rubber Exchange, is in Cereal House, Mark Lane. Trades represented include cocoa, coffee, soya bean, rubber, sugar, oil and wool.

Bury Court House

This is a major new office building completed 1983 in the shipping, banking and insurance sector of the City of London. The development has frontage to Bevis Marks and adjoins the Baltic Exchange Building just a short distance from the Lloyds Building and Liverpool Street Station. On each upper floor there are screened and planted balconies which widen at each level and overhang the ground reception with its sculpture.

The Monument, Billingsgate and Custom House

Cannon Street Station
Part of London's Roman past was unearthed during redevelopment of Cannon Street Station in the early 1980s. The oak piling - possibly the earliest example of small displacement piles - support a stone wall in silty sloping soil. It is possible that it formed part of a Roman Governor's Palace in London.

London Stone
The Roman Milliarium is the stone marking the point from which all road distances were measured. It is set into the wall of the Bank of China opposite Cannon Street Station in the City of London.

Billingsgate New Office Blocks
The development comprises two new interconnecting office buildings on the former lorry park site adjacent to the historic fish market. The glass clad towers cascade from ten storeys to the west through a series of stepped roof terraces, to three storeys on the boundary of the fish market building. During the redevelopment of the site, the excavations uncovered the plan and walls of a vanished medieval church of St Botolph and its 15th century vestry, as well as wooden revetments or quays of the waterfront at different periods were discovered.

Old Billingsgate Market
Billingsgate in Lower Thames Street, EC3, was London's old fish market and dates back to 1876. As part of the Port of London, it received the bulk of the nation's fish. It was famous for its porters with their unique leather hats on which they carried fish boxes and their use of strong Eastender's language. The market was moved in 1983 to the North Quay in the Isle of Dogs.

Custom House
Adjacent to the old Billingsgate building, the Custom House was built for the Port of London's administration building in 1828 to the design of Robert Smirke. Badly damaged during the Second World War, almost half of it was rebuilt. The Portland stone frontage to the Thames is highly imposing.

The Monument
The Monument in Monument Street, EC3, stands 200 yards from the site of the baker's premises where the Great Fire of London started on a Saturday in September 1666 and continued till late on the following Wednesday night with considerable destruction of houses, churches, Livery Halls and warehouses. The fine 17th century hollow fluted column, 62m (202ft) high, was designed by Wren and has excellent views of London. Unfortunately it is closed to the public.

Trinity House
River Pilots took over inward-bound ships from the Channel Pilots at Gravesend and bought the ships up to the former dock entrances or wharves in London. The Pilots were trained and licensed by the Corporation of Trinity House and had to hold foreign-going masters' certificates.

Former PLA Building
This Grade 2 listed building in Trinity Square, EC3, was designed by Sir Edwin Cooper in 1913, and was opened in 1922. It was the headquarters of the Port of London Authority until it was sold in 1972.

Corn Exchange
Mark Lane, EC4. The Exchange is the largest market in the country for all kinds of cereals, flour, agricultural seeds, animals foodstuffs, hay and straw, seed potatoes etc. It has a membership of 400 firms, approximately 100 of which have stands at the floor.

Fur Market and Beaver House
The London Fur Market comprises two fur auction companies, a large number of fur merchants and commission agents whose offices are located around Beaver House. This is the largest fur market and auction centre in the world.

Billingsgate and Custom House

Map of City waterfront

Scale:

0   1/4   1/2 Mile

0   1/2   1 Km

53

City Heritage Walk Around Bank of England

The following walk starting from Bank Station takes about an hour and covers one and a half miles.

Bank of England and Museum (1) Founded in 1694 to finance the war against Louis XIV of France, it has since become the bank of government and the nation. The present building, designed by Sir Herbert Baker, was completed in 1937. The adjacent museum tells the story of the bank.

Royal Exchange (2) Founded by Sir Thomas Gresham in 1566, the present building was designed and completed by Sir William Tite in 1844. Used as an exchange for merchants for four centuries, it is now occupied mainly by insurance organisations.

Building 33-35 Cornhill (3) Ornate classical building of Portland stone completed in 1857.

Simpson's Tavern (4) An old tavern located in an area of alleyways and courts.

Church of St Michael (5) Built by Wren 1670-2, except for the tower.

Church of St Peter (6) By Wren 1677-81, the churchyard contains two London plane trees.

Building 140-144 Leadenhall Street (7) Designed by Lutyens, the Portland stone building dates to 1929.

P & O, and Commercial Union (8) The two modern buildings by Melvin Ward were built 1964-9.

Church of St Helen (9) It is the remains of a Benedictine nunnery.

Church of St Andrew Undershaft (10) The name comes from the shaft of the maypole. The tower and church were mainly built during 15th and 16th century.

Lloyd's Building (11) The international and world famous insurance market, the new building on the east side of Lime Street, was erected by Heysham 1950-57 and is connected by a bridge to the old headquarters by Sir Edwin Cooper 1925-28.

Office Block 34-36 Lime Street (12) The building by Sheppard and Robson was completed 1974.

Leadenhall Market (13) The Victorian retail market by Sir Horace Jones 1881 has many shops devoted to food and poultry trade. Beneath the market archaeologists found the site of the Roman Basilica.

Building 2-3 Philpot Lane (14) In the Court at the rear of No.1, you find a much altered 18th century building.

Building 4,7, and 18 Philpot Lane (15) This is a late 17th century four storey building.

Building 23-25 Eastcheap (16) A mid 19th century corner building of polychrome brick.

Building 33-35 Eastcheap (17) A gothic design built in 1877.

Church of St Mary-at-Hill (18) Built by Wren 1670-76 to replace an older church first mentioned in 1177.

One Tree Park (19) The area is frequented by traders.

The Monument (20) Designed by Wren, it was erected 1671-77 to commemorate the Great Fire of London of 1666. The height of 200 ft is said to be the distance westward from the baker's shop in Pudding Lane where the fire started.

Church of St Edmund the King (21) Built by Wren 1670-79.

Building 60-62 Lombard Street (22) Early 20th century construction.

Church of St Mary Woolnoth (23) Built by Hawksmoore 1716-27.

Building 1 Cornhill (24) Recent design, recent building of Portland stone with rounded corner supporting a dome.

Mansion House (25) Official residence of the Lord Mayor of London during his year of office. It was completed in 1753.

Chapter House (26) Designed by Wren 1712-12, gutted in the last war but now restored.

Leadenhall Market

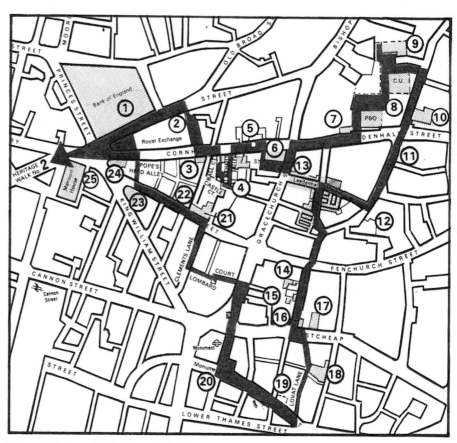

Map of Heritage Walk Around Bank

City Heritage Walk Around St Paul's

The Monument

Paternoster Precinct (27) Developed 1961-67 following extensive war damage.

Tower & remains of Christchurch (28) On the site of the Chancel of the Franciscan (Greyfriars) Church.

National Postal Museum (29) Contains Phillips collection of 19th century British Postal stamps and registration sheets 1840.

Statue of Rowland Hill (30) Founder of penny post, by Onslow Ford 1881.

Postman's Park (31) Contains Minotaur sculpture by Michael Ayrton, and a memorial shelter.

Church of St. Botolph, Aldersgate (32) Founded before 1291, and rebuilt 1788-91.

Gateway Leading into Churchyard (33) Has picturesque upper storeys.

Church of St Bartholomew-the-Great (34) In 1123 Rahere, a prebendary of St Pauls, founded St Bartholomew's as a priory.

Drinking Fountain (35) Bronze female figure by J.B. Philip dated 1873.

43-45 Cloth Fair (36) Late 18th century houses and shop fronts.

41-42 Cloth Fair (37) Much restored late 17th century houses.

Smithfield Market (38) Designed by Sir Horace Jones circa 1868 and owned by the Corporation of London.

Hand & Shears Public House (39) Early 19th century corner building.

Museum of London (40) Designed by Powell and Moya 1974 houses the collections of the London and Guildhall museums.

Barbican (41) The centre for Arts and Conferences, the largest in Western Europe. Barbican means a projecting watch tower over the gate of a fortified town.

Roman Wall (42) Sections of the Wall still survive on the north side of London Wall.

Church of St Giles Cripplegate (44) It stands in the middle of the Barbican area and was restored after war damage.

Upper level walkway (45) A link to major centres in the City.

Exhibition Hall (46) An exhibition featuring aspects of City life.

Sculpture (47) Beyond Tomorrow by Karin Jonzen, 1972.

Glass Fountain by Alan David (48).

Stainless Steel Sculpture (49) Called Ritual by Antanas Brazdys, 1969.

Guildhall (50) Centre of the City's government for more than 1,000 years.

Church of St Lawrence Jewry (51) 1671-87 by Wren. Restored in 1954-57 after Second World War by Cecil Brown.

Irish Chamber (52) Early 19th century corner building of yellow brick and stucco.

42-44 Gresham Street (53) Italianate corner building of Portland stone circa 1850.

3 King Street (52) Distinguished classical design by Thomas Hopper in 1836.

Church of St Mary-le-Bow (55) Another Wren church built 1670-83.

6-8 Bow Lane (56) Mid 19th century commercial building of polychrome brick.

24-26 Watling Street (57) Mid 19th century commercial building.

19-21 Watling Street (58) Mid 19th century building of yellow brick and stone.

Festival Gardens (59) Constructed at the time of the Festival of Britain in 1951.

Tower of Church of St Augustine (60) 1680-83 by Wren. The graceful spire has been reconstructed following war damage.

St Paul's Cathedral Choir School (61) Designed by Leo de Syllas of the Architects' Co-Partnership, 1962-7.

St Paul's Cross (62) Rebuilt 1910 to mark the site of a preaching cross.

Church of St Mary Aldermary (63) Wren rebuilt the Church in 1682 after the Great Fire.

Remains of the Temple of Mithras(64) Foundations of a temple to a Roman Sun God were discovered in 1954.

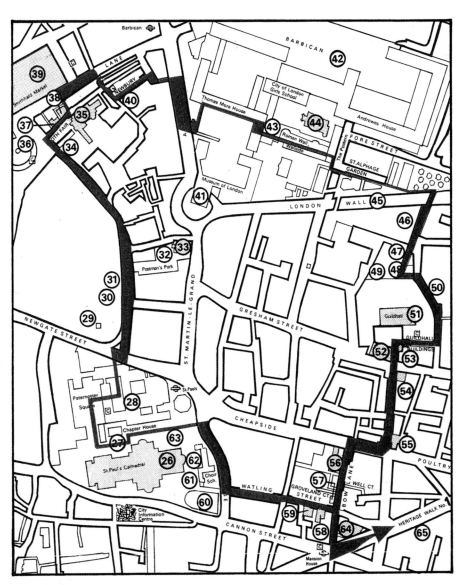

Map of Heritage Walk around St Paul's

Tower of London and The Crown Jewels

The Imperial State Crown

Tower of London

The 900 year old medieval fortress has over 3 million visitors a year. The ravens, Beefeaters, armouries and the well-protected Crown Jewels are among the great attractions. It is today one of the top ten tourists attractions in the capital. A few centuries ago, visitors were not admitted into the interior as it enclosed an arsenal of guns and ammunition for the royal ships and contained a prison for traitors to the Crown. The conditions in the prison's dark cells were appalling. There were other prisons in London, including the Clink in Southwark but conditions were not as grim as in the Tower.

The White Tower

As a royal palace and fortress, the White Tower was built 1078-1097 to accommodate the King on the upper floors and the Constable of the Tower, who commanded the garrison, at the lower level. The present top floor was built much later. The basement contained the storerooms and the well. Later the building was used as a prison, an armoury and a storehouse for royal costumes and furnishings. Currently it houses the historic Royal Armouries.

The building rises 90 feet (27.4.m) to the battlements and in plan measures 118 feet (35.9.m) by 107 feet (32.6.m). The walls are 15 feet (4.6.m) thick at the base and 11 feet (3.3.m) at the top. They were built of Kentish ragstone, quarried near Maidstone, and whitewashed regularly, hence the name White Tower. The fine Chapel of St John the Evangelist was built of limestone imported from Normandy and it is still used occasionally for worship.

The Yeoman Warders

The Yeoman Warders of the Tower and the Yeomen of the Guard, popularly known as 'Beefeaters', provide the guard at the Tower. The Ceremony of the keys, after the locking of the gates, is a colourful event.

The Bloody Tower

Previously controlled the watergate, the first floor contains the windlass that still operates the portcullis at the front of the gatehall below. The tower was used to accommodate such eminent prisoners such as Archbishop William Lund (in 1640-45) and Lord Chancellor, Jeffreys of the 'Bloody Assizes' (in 1688-89).

The Crown Jewels

The Jewel House contains the coronation ornaments and robes, a number of historic crowns, state swords, banqueting and church plate, insignia of the orders of chivalry and medals. The Imperial State Crown, worn by the monarch at major state occasions, is encrusted with more than 2,800 small diamonds as well as the Second Star of Africa, one of the nine major stones cut from the Cullinan Diamond. The beautiful St Edward's Crown, which holds the legendary Koh-i-noor diamond, is used only at a coronation. The head of the Sceptre with the Cross contains the Star of Africa, at 530 carats the largest cut diamond in the world. Changes are being made so that the Crown Jewels are displayed with more circulation area while the armour is being moved to a new museum in Leeds Castle. The Tower of London will retain the Armouries relating to its own history.

The Royal Armouries

The collection contains mainly the Tudor and Stuart royal armour including the relics of Henry VIII's personal armoury. There is armour for an Indian elephant, probably the largest in the world, said to have been captured at the battle of Plassey in 1757 when Robert Clive and troops of the British East India Company defeated the army of Suraj-ud-Dowlah, the Nawab of Bengal.

The White Tower

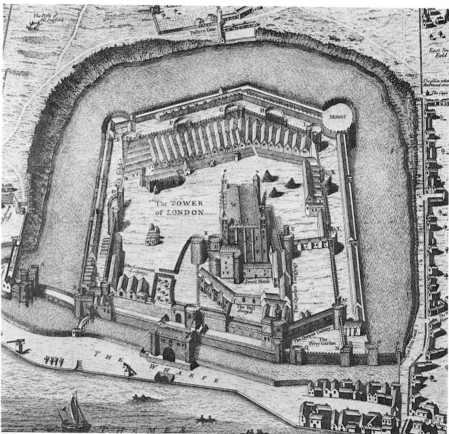

Tower of London, circa 1597

Tower of London

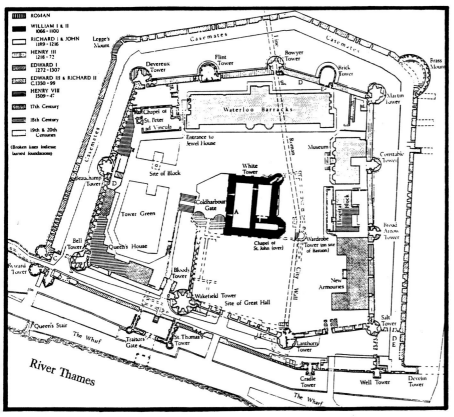

Map of Tower of London

Queen's House

The charming row of Tudor houses, circa 1540, are located in the south west corner of the Tower. They are the last of the remaining half-timber houses which existed in the City of London before the destruction by the Great Fire of 1666.

The Tower Ravens

Visitors to the Tower are always curious about the large black ravens that hop and jostle about in the grounds. They are protected by ancient tradition and are looked after by an officer of the Crown. Legend has it that "If no ravens are in the Tower, then the White Tower will fall and the whole British Empire will collapse!" The tradition may have originated at the end of the 11th century when the Tower was completed. The first residents threw their kitchen refuse out of the window slits onto the ground beneath; the ravens would fly down from the nearby woods to feast on it! Today, each morning, the Raven Master lets out his charges and gives them water and horse meat. Since 1835 they have had one wing clipped to prevent them flying away. They accept scraps and buns from the daily sightseers.

Tower Bridge, Museum and Tower Hill Pageant

Tower Bridge

Tower Bridge has been for over 100 years and remains a symbol of London and one of the world's most famous landmarks. The foundation stone for the bridge was laid on 21st June 1886, and the ceremony was performed in the presence of thousands of spectators, by Edward, Prince of Wales, who later became King Edward VII. The event led to eight years of intensive construction work under the direction of the engineer Sir John Wolfe-Barry and the architect Sir Horace Jones. In June 1894, the bridge was opened providing a spectacular crossing of the River Thames. In its first month of operation, Tower Bridge's steam engines raised the twin roadway bascules 655 times to allow ships access to the busy riverside wharves in the Upper Pool between London Bridge and Tower Bridge. Nowadays, with the move of the Port of London downstream to Tilbury, they are in operation no more than three or four times a week. This is often for the passage of naval vessels bound for moorings alongside the museum ship HMS Belfast moored off London Bridge.

Tower Bridge is one of London's greatest attractions and during the tourist season attracts over 2000 visitors a day. They are able to take a high-speed lift up the north tower to the enclosed high, glazed, walkway, which has port holes for taking photographs.

Enjoy and admire a unique panorama of London and the Thames from the 139 feet high overhead walkway!

Tower Bridge Museum

Visitors can cross the walkway and go down via the south tower where there is a museum containing exhibitions illustrating the history of the bridge. There is also a tour of the old Victorian pumping engines, the control room with its array of levers and signal bells which controlled and provided the power to raise the bascules before its conversion to electricity. The intricate workings of Tower Bridge are explained by "animatronic" people and videos in a new exhibition. There is a bookshop in the museum.

All Hallows by the Tower Church

This church previously known as All Hallows, Barking was a foundation from Barking Abbey, circa 660-70 shortly after the Abbey, its mother house, was completed. During the second world war much of the church (mainly the nave) was destroyed by bombing. This, however, conveniently revealed the foundations of a Saxon church circa 7th-8th century, as well as traces of a Roman temple dedicated to Mithras. The church's aisle is of Norman origin as was the original chapel and crypt, before being replaced in the 14th century. The west end tower was built between 1658-9 although the spire was built late last century. The vestry was built 1931-2 and the earlier part which was bomb damaged, was rebuilt in 1949.

Church Shipping Heritage

The church retains its links with the Port of London and London's shipping companies, whose coats of arm feature in the South Aisle windows. The Mariner's Chapel contains models of ships given to the church in thanksgiving and a crucifix made of wood from the Cutty Sark.

Tower Hill Pageant

Situated opposite the Tower of London, this is London's first dark ride museum with an imaginative display to give an insight into the capital's 2000 years of history. The audio/visual and shopping complex are beneath the streets of London. Automated cars travel in time, taking visitors through scenes depicting the historical development from the Roman settlement in 43AD to the modern time. You can see life-size scenes of the Romans, Saxons and Vikings. Archaeological displays of items discovered along the Thames help to complete the story of the Port of London. The vaults were originally the basements of the Mazawattee Warehouse, built in 1864 by tea merchants. The upper floors were destroyed during the Blitz of 1940 and the surviving vaults were adapted for the museum in 1987.

Tower Bridge

CENTRAL LONDON SOUTH OF THE RIVER THAMES

South Bank Arts Centre

New Flower Market and Battersea Leisure Park

New Covent Garden Market

In 1838 the London and Southampton Railway terminated at Nine Elms, north of Battersea Park, and was extended to Waterloo ten years later. For over a century it functioned as a large rail freight shunting station for the Port of London until the port activities were moved to Tilbury in the 1970's and the docks in London were closed. The site was subsequently developed as the new Covent Garden Market for supplying London with fruit and vegetables. The market is a steel space frame building by the Thames, at the south east corner of Vauxhall Bridge. Square in plan, the roof floats above the 36 tree-like support stanchions. It is designed to reflect the delicate nature of the flower trade it encloses. A repetitive pattern of translucent roof lights and the diagonally set space frame is said to create an internal effect as of trees and intertwining overhead branches. The traders have their offices on a mezzanine floor around the perimeter of the market. It is packed in the early mornings.

Albert Bridge

This is a suspension bridge supported by two turreted arches made of cast iron. The bridge was designed by R. Ordish and built 1871-7.

Battersea Park and Pagoda

The park covers about 200 acres including a 15 acre lake on which boats are available for hire. There are also a children's zoo, adventure playgrounds and an athletics track. All-weather pitches and a bowls club are added attractions. It is the setting of the London Peace Pagoda erected in 1985. The Park itself has recently been improved, its cascades restored, railways renewed, new gardens and facilities added. It is located between Albert Bridge Road and Queenstown Road, SW11 (081 871 7530).

Battersea Power Station

The disused Power Station is situated on the banks of the Thames to the east of Battersea Park. Built 1932-34 to supply London with electricity, it was designed by Sir Gilbert Scott, the architect for Waterloo Bridge. The Grade 2 listed building has been the subject of several proposals for leisure facilities.

Battersea Dogs Home

One of Wandsworth's most famous places, the Home draws more than 60,000 visitors a year to see stray animals rescued from all parts of London. This is three times as many visitors as there are dogs to see. The Home estimates that it takes nine visitors to secure a new home for one dog or cat. It has been operating since 1860 and its excellent work must be supported by the public.

Battersea Bridge

Completed in 1890, the bridge was designed by Sir Joseph Bazalgette and his son Edmund, the former being the Chief Engineer of Metropolitan Works. It consists of five steel arches bearing on granite piers. The arches and piers have been integrated into a well proportioned structure decorated with oriental motifs. The fine bridge is a Grade 2 listed structure.

Plantation Wharf

Plantation Wharf, on the south side of the River Thames between Battersea and Wandsworth, is an example of a 1980's empty office block that has been converted into private homes. On the site of a former candle factory, the 12-storey tower was recently transformed into more than 50 apartments. The floor-to-ceiling windows provide an abundance of natural light. There is an opinion among some people commuting from the suburbs that they would rather live and work in the City and believe converted offices could become the mansion blocks of the 21st century!

Clapham Junction

Britain's largest railway junction at Clapham lies in the Borough of Wandsworth. The station was built in 1845 linked to Waterloo three years later, and served what was already a rapidly developing area of factories and workmen's dwellings which replaced the market gardens in the area.

The Japanese Pagoda, Battersea Park

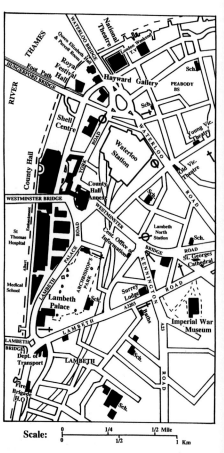

Map of South Bank

Lambeth Palace, County Hall and War Museum

The County Hall

Lambeth Palace

The medieval palace has been the London seat of the Archbishop of Canterbury since 1197. The Gothic style hall of residence was built during the 19th century. The crypt under the chapel is the oldest part of the building. The chapel dates to the middle of the 13th century and was restored following bomb damage during World War II. The gatehouse, Morton's Tower, was completed in 1490. The Great Hall built 1663, contains the fine library with over 100,000 books and rare manuscripts, including Elizabeth I's prayer book. Every ten years the Assembly of Anglican Bishops from all parts of the world is held here. The Great Hall contains portraits from the 16th to the 19th century.

Museum of Garden History

Housed in a church close to Lambeth Palace, the museum commemorates John Tradescant and his son, gardeners to Charles I. The churchyard has the grave of Captain Bligh of Mutiny on the Bounty story.

County Hall

The London County Council was formed in 1889 to supersede the Metropolitan Board of Work, providing the capital with a system of local government which lasted for nearly one century until the demise of the Greater London Council in March 1986. Lambeth Council has recently given planning permission for the conversion of the main block of the historic building into a 600 bedroom hotel. The Japanese Group who purchased the building for £65m intend to restore the building to its original oak-panelled splendour of 1922, designed by the architect Ralph Knott. The proposed name for the hotel is The County Hall, with the old Council Chamber becoming a conference centre. A small wedding chapel, health club and a riverside restaurant are some of the features being planned.

Lambeth Bridge

The bridge, completed 1932, consists of five segmental steel arches with bearings on granite piers. It is ornamented with pierced parapets, handsome iron lamp standards and granite obelisks on the four corners. The engineer was Sir George Humphries and the architect Sir Reginald Bloomfield.

International Maritime Organisation

The IMO offices are on the Albert Embankment on the south side of the River Thames upstream of Lambeth Bridge. The International Maritime Organisation is the only United Nations organisation based in the United Kingdom.

Vauxhall Bridge

Designed by M Fitzmaurice and opened in 1906, the bridge has a steel superstructure and the abutments are faced with granite. It replaced the first iron bridge across the Thames built in 1816.

Vauxhall Cross

An office complex has recently been completed on the south eastern end of Vauxhall Bridge. The accommodation is in a progressively stepped configuration, set back from the River Thames in a group of three longitudinal blocks linked by glazed courtyards and atria. It is occupied by the Secret Intelligence Service MI6.

St. Thomas's Hospital

Founded in 1106, the Medical School was opened in 1871. The present modern building overlooks the river and the Houses of Parliament.

Imperial War Museum

Located in Lambeth Road, SE1, the museum contains exhibits demonstrating the development of warfare and over three million photographs of the world wars since 1914. There are over 25 million feet of film dealing with all aspects of warfare. It houses the Foreign Documents Centre of other foreign war archives. Other exhibits include a Second World War Spitfire, a small ship from Dunkirk, the original surrender document of the German Army in 1945 and the headgear and rifle of Lawrence of Arabia. There are also film shows (071-416 5000).

St. George's R.C. Cathedral

First mentioned in 1122 and rebuilt 1510-27, it was enlarged in 1629 and further rebuilt by John Price during 1734-6.

Cuming Museum

The archaeological and local history museum which Richard Cuming began in 1782, is located at 155 Walworth Road.

The Oval Cricket Ground

Dating from 1846 the Oval is the headquarters of the Surrey Cricket Club. Test Matches are played on this ground and the first Test Match between England v Australia was played here in 1880.

Floating Fire Station

At the Albert Embankment, the new floating fire station is a steel double decker pontoon providing dormitories and other facilities.

Royal Festival Hall & South Bank Arts Centre

The Royal Festival Hall

The South Bank Arts Centre
London's South Bank has one of the world's greatest concentrations of year round cultural attractions, most of them closely situated between Westminster and Waterloo Bridges and within easy walking distance of Waterloo Station. The galaxy includes the Royal Festival Hall and Queen Elizabeth Concert Hall, the National Theatre with its three auditoria, the Hayward Gallery with two major exhibitions in progress, and the National Film Theatre.

South Bank Exhibition
The Festival of Britain in 1951, held at the South Bank, 100 years after the Great Exhibition in Hyde Park, was meant to show the contributions made by Britain to civilisation by advances in science, technology and industrial design. A striking feature was the Dome of Discovery containing exhibits on modern exploration.

Royal Festival Hall
Opened for the Festival of Britain 1951, it has a huge stage and superb acoustics with accommodation for 3000 people. Part of the South Bank Arts Centre, it hosts a variety of concerts and theatrical events.

Hayward Art Gallery
Part of the Queen Elizabeth Hall complex on the South Bank, it provides a series of exhibition galleries at different levels, together with an adjoining sculpture hall.

Royal National Theatre
It is well known for its English plays, including MacBeth by William Shakespeare and An Inspector Calls by J B Priestley. The building was designed by Sir Denys Lasdun and opened in 1976 (071 633 0880).

Museum of the Moving Image
Completed in 1989 this museum is tucked away under the arches of Waterloo Bridge. It packs an amazing amount of information in what appears to be a small space from the outside. It is said to be the world's largest museum devoted to cinema and television. Its technology includes more than 70 computer-controlled video screens.

Florence Nightingale Museum
A short distance west, past the Royal Festival Hall, the former County Hall and Westminster Bridge is another attraction opened in 1989, the new Florence Nightingale Museum. It is by the vehicular entrance to St. Thomas' Hospital which is on the south bank of the river opposite the Houses of Parliament. It is a monument to one of the greatest and most influential personalities of the 19th century.

London Weekend Television
Headquarters of one of the largest independent TV companies, the week-end news summaries are relayed from this studio.

Waterloo Bridge
A modern concrete bridge built 1942, and faced with Portland stone. The architect was Sir Giles Gilbert Scott and the consulting engineers were Rendel, Palmer and Tritton, who designed Chelsea Bridge. There are two ranges of segmental arches with massive concrete beams spanning between them.

Shell Building
This tall building is the London Headquarters of Shell International, the largest British oil company and also one of London's highest viewpoints.

Stamford Wharf and Oxo Tower
Next to the South Bank Festival Complex is the Coin Street Community development where a number of buildings are under renovation. The structural repair of Stamford Wharf and the familiar Oxo Tower were completed in 1992 and the refurbishment involves the provision of apartments, shops, restaurants and a River Thames Centre. There is also a proposal to add a helicopter pad on the roof of the adjacent massive riverside Sea Containers House.

Bankside and Shakespeare's Globe Theatre

Bankside and Southwark Riverside

Southwark, the Docklands area of London on the south side of the River Thames, was a bustling maritime community from the Roman times. The Cathedral near London Bridge dates back to the 13th century and is rich in architectural features and memorials. To the west is Bankside, which in Elizabethan times was the 'Broadway' of London with pubs and theatres and previously it was known as the 'clink' because of the nearby prison. Not being under the jurisdiction of the City of London, the district became a refuge for law breakers and a place with bear pits and bull rings! At the same time it developed the fine Globe Theatre where shakespeare's plays were performed.

The Southwark riverside walk has some of the finest views of the City and St. Paul's Cathedral just across the Thames. The site of Shakespeare's Globe Theatre is marked by a plaque in Park Street. A replica of this theatre is being completed nearby. It is said that Sir Christopher Wren stayed at No.41 while St. Paul's was being built.

Shakespeare's Memorials

During the Middle Ages, the area around Southwark Cathedral was known for its theatres and playhouses. Shakespeare's Globe Theatre was built of timber in the area and there were others at Bankside. A modern memorial window in the south aisle of Southwark Cathedral commemorates William Shakespeare and shows characters from his plays. There is also an alabaster figure of Shakespeare carved by Henry McCarthy in 1912 with a description of 17th century Southwark which includes the Globe Theatre.

Shakespeare Theatre Museum

Located at Bear Gardens, Bankside, SE1, the museum traces the development of the theatre in London from 1576 to 1642 underlining the history of Southwark as an important theatre area and being the home of Shakespeare's Globe Theatre. Text panels and a replica of the 1616 stage tells the story of old English theatre (071 261 1363).

New Globe Theatre

Built originally in 1599, Shakespeare's Globe Theatre was destroyed by fire in 1613, rebuilt in 1614 and finally demolished 1644. After nearly 400 years, the Globe Theatre is being reconstructed and can be seen from the River Thames. The sections of the two three-tier seating bays of the replica theatre have been completed. Master craftsman used 200 year old oak beams and traditional 16th century building techniques for timber framed buildings. Its site is near to where archaeologists found remains of the original theatre when excavations were made for a new office block. Scheduled to be completed in 1994, the International Globe Centre will consist of a theatre, restaurant and museum. American actor, Sam Wanamaker received a civic honour from the Queen for his sponsorship. The theatre is intended as a forum for educational purposes and will help teachers. There will be a permanent exhibition, library and a collection of audio-visual performances of Shakespeare's plays. The aim is to establish a Globe repertory company. The American sponsored Trust has a 125-year lease on the site and the building was completed in 1994.

Anchor Inn on Bankside

The original pub on the site was destroyed in the Great Fire of London in 1666. It was the haunt of river pirates and the warders from the nearby prison. The present building dates from the 18th century and is frequented today by Londoners working nearby and visiting tourists. A copy of Dr Johnson's first dictionary is held by the pub.

The George Inn

The George, a galleried coaching inn, is the last remaining of its kind. The old interior is well preserved and to go inside is to step back into history. Shakespeare performed his plays here.

Bankside Power Station

The Bankside Power Station on the South bank between Blackfriars Bridge and Southwark Bridge was operational between 1963 and 1979 and closed in 1981. It is proposed to demolish the building for redevelopment possibly a national gallery.

William Shakespeare

A model of the re-constructed Shakespeare Globe Theatre

Historic Southwark Cathedral

Southwark Cathedral

Situated at the south western end of London Bridge, Southwark Cathedral is one of several great cathedrals in London. The first church was built during the 7th century on the site of a Roman temple. There are Roman tiles in the pavement at the entrance to the south choir aisle of the present building. Early in the 12th century, a Norman priory church was erected and called St. Mary Overie which meant 'over the river'. Some traces of this church still survive in the south aisle and an arch in the doorway of the nave's north aisle.

Dating from about 1275, there is an oak effigy of a knight with his ankles crossed and one hand on his sword pommel. The crossing under the central tower has four piers dating from the 14th century. Carved on the east wall in the south transept are the hat and coat of arms of Cardinal Henry Beaufort, Bishop of Winchester, who helped to finance restoration following fire damage in 1385. Following the Reformation in the reign of Henry VIII, the Church Priory was suppressed and changed its name to the Church of St. Saviour in Southwark.

The birth in 1607 of John Harvard, who founded Harvard University in America, and his baptism at St. Saviour's, is commemorated by Harvard Chapel. The present tower with four pinnacles was completed in 1689. The massive pulpit was erected in 1703. The Prince of Wales laid the foundation stone for the nave in 1890 and in 1897 St. Saviour's became the Cathedral of South London, Southwark. Today it has been restored to its full glory. The Augustinian priory of St. Mary Overie was founded on this site in 1106AD. The cathedral has a vaulted Quire and triforium and a four aisled retro-quire all of Early English design. Since the early 19th century the nave has been re-built twice; latterly by Bloomfield in 1890 to 1897. The exquisite lady chapel was demolished c1830 to make way for a new London Bridge.

Southwark Bridge

The bridge was designed by Sir Ernest George and opened by King George V in 1921. The old bridge, erected by a private company, opened in 1819. It was purchased by the City of London in 1866 and the toll for crossing was abolished.

Southwark Roman Catholic Cathedral

Designed by A. Pugin and built between 1840 and 1848. The cathedral had its design drastically altered after extensive bombing in 1941.

Gabriel's Wharf

Gabriel's Wharf is an excellent example of a recent conversion of derelict riverside buildings on the South Bank near the Festival Hall into craft workshops, market stalls, and restaurants. There are some 20 workshops where artists, designers and craft people produce their goods and visitors can stroll around and watch them. The crafts include china, rugs, jewellery, woodwork and sculpture. Most of the work is for sale and special commissions are undertaken. It is within a short walk of Blackfriars underground station.

Southwark Cathedral and riverside, circa 1820

Hays Galleria and HMS Belfast Museum

London Bridge City

The Pool of London, situated between London Bridge and Tower Bridge, was for centuries one of the best known parts of the River Thames. Crowded with shipping of all shapes and sizes and lined by wharves and tall warehouses, huge quantities of cargo were handled there each year. Today, all that is gone and large cargo ships no longer venture into London. The wharves have been closed and the warehouses have been converted, whilst keeping their original facades, into fashionable shops and offices. This is clearly seen on the south side of the river where the new buildings of London Bridge were recently erected. The old "London Larder" of Hays Wharves, where ships once anchored to unload tea, coffee, spices and liquors from all over the world, is now one of the most exciting riverside shopping and eating places in London.

Hays Galleria

One of the developments of London Bridge City completed in 1988, the beautiful Hays Galleria shopping precinct was previously Hays Dock, built by Cubitt in 1856 around the River Neckinger which flowed into the Thames. This accounts for the bend in the building which provides so much of its old character. The shops at ground level create a street in the tradition of the Victorian arcade. The design was partly inspired by the Galleria Victor Emmanuel in Milan, Italy. Hays Galleria links Tooley Street with the new river walk, previously not open to the public. The glazed roof provides a magnificent space in which people can meet, stroll, shop and eat by the river and be independent of the weather conditions.

There are colourful craft stalls and a piece of sculpture, The Navigators, which echoes the wharf's maritime past. From the riverside walkway you can view and photograph the contrasting skyline of old and new London on the north bank of the Thames. The Horniman Pub and Restaurant has terrace seating overlooking the river.

London Bridge

This three span prestressed concrete bridge, constructed 1968-70, is the most recent of London bridges, and it replaced the old stone London Bridge of 1831. It is 262 metres (860 feet) long and consists of four parallel concrete box beams forming three arches resting on concrete piers and abutments. The central span is 103 metres and the two side spans are each 80 metres. The bridge piers have a facing of axed granite and the parapets have a polished granite finish, otherwise the superstructure presents its Creetown granite aggregate concrete surface

as precast. The pier positions were planned to meet the requirements for navigation widths. The consulting engineers, Mott, Hay and Anderson were responsible for the design. The concrete casting yard was laid out at the old Russia Dock in the Surrey Docks and the old dock was filled in with spoil excavated from the bridge foundations.

In April 1968 the old bridge was sold to a Californian property company for just over £1 million and its numbered and identified stone blocks were shipped to the United States. The bridge was re-erected in 1971 at the holiday resort of Lake Havasu City, Arizona. During the same year Disney Land was also opened in California.

HMS Belfast

HMS Belfast is the first warship since Nelson's ship HMS Victory to be preserved for the nation. It is the last survivor of the Royal Navy's big gun ships and is permanently moored in the River Thames off London Bridge as a floating museum. Launched in Belfast in 1938, she was damaged by magnetic mines in 1939 at the outbreak of World War II and re-entered

service in 1942. She played a key role during 1943 in the Battle of North Cape which resulted in the sinking of the Scharnhorst. The ship was converted into a museum ship in 1971.

A marked route shows visitors around the ship. The areas to be discovered include the Bridge, the mess deck, the Boiler and Engine Rooms, the Punishment Cells and two 6 inch gun turrets. There are special displays about D Day, the development of warships gunnery and other naval matters. The excellent museum has an educational service for schools and colleges. London Bridge station is within a few minutes walk.

London Dungeon

The world's first horror museum has a series of life-size tableaux of gruesome scenes and is not recommended for children under the age of 10. The exhibitions of medieval history show superstitions, witchcraft, punishment and death. It provides an insight into the horrors of murder and torture, which are illustrated here in the dark vaults under London Bridge station in Tooley Street, SE1 (071 403 0606).

Hays Galleria, London Bridge City

EAST LONDON AND LONDON DOCKLANDS

St Katharine Docks and City of London

From the borders of the City of London eastwards, London Docklands have changed over the past two decades. Traditionally, a working class area associated with the docks and Port of London, new developments have bought about many changes. St Katherine's is a great business and tourist centre with a beautiful marina and yacht club. Petticoat Lane Market is still thriving but the main new attraction is Canary Wharf with Britain's tallest tower on the Isle of Dogs. Most of Britain's national newspapers are now printed in Docklands. Arrive in East London Docklands and you will find a city reborn into a new era but with a unique heritage.

Some present day East Enders are descendants of the legendary dockers who worked for nearly two centuries in the greatest complex of docks in the world. They were recognised for two things - their friendliness and the Cockney accent. It is no coincidence that one of Britain's top television series, East Enders, is watched and loved all over the world. A more recent phenomenon is the influx of new people into the area - in the form of business people and young city dealers who want to live in the quaint converted warehouses and the exciting new riverside apartments and luxury waterside homes.

St. Katharine Docks and Petticoat Lane Market

Wapping and Limehouse

Starting at Tower Bridge, the western gateway to Docklands, there is the haven of St. Katharine's Docks which was opened in 1828 and is currently one of the most popular tourists' centres in London. Further east is Wapping and Shadwell Basin where the London Docks were built in 1804. The Basin was built in 1860 as an extension to these docks but is now surrounded by new housing and a sailing club for the young EastEnders. A short distance away is the Limehouse Basin, otherwise known as the Regents Canal Dock which linked the River Thames and the East London Docks with the Grand Union Canal and other inland waterways. The latest Limehouse Link road tunnel passes underneath the north side of the basin.

St Katharine Haven

If you are looking for mooring facilities in the City of London, you can set sail for the established St Katharine Haven, 43 nautical miles upstream of Sea Reach No 1 Buoy. The Haven is also the headquarters of the Cruising Association with its library and has a number of visiting Thames sailing barges. The vessels include 'Lady Daphne' which is used for charter, the famous 'Nore' lightships, the S.V. 'Yarmouth' an 1859 steam ferry and the S.T. 'Challenge', the last steam tug to work on the Thames.

Royal Mint Court and Site

In 1989 the Queen opened the new eight storey office building, Royal Mint Court, opposite St Katharine's Docks. She was presented with a set of commemorative sovereigns since 1989 was the 500th anniversary of the first minting of the coin. Archaeological work undertaken between 1986-88 uncovered evidence of the long history of the site. In the Middle Ages, the site was a Black Death Cemetery (1349) and the location of St Mary Graces Abbey (1350-1536). The Royal Mint itself was located here from 1806-1967 before being moved to Wales.

Petticoat Lane Market

Close to Liverpool Street Station, E1, the Sunday market dates back to the 17th century when stalls for selling clothes were first established here. Today, it contains stalls for clothes, jewellery and household goods and is packed with Londoners and visitors each week. Nearby streets have developed into centres for furniture, electrical goods and pet animals at Houndsditch Sunday market.

Spitalfields and Market

Spitalfields on the eastern boundary of the City of London was renowned for its famous vegetable market and clothing industry. Many of the former merchants' houses in this area, occupied by the immigrant Huguenots in the 18th and 19th centuries, have been restored to their former glory. In 1987 the City of London Corporation, invited developers to relocate the vegetable market at Temple Mills and produce plans for a mixed scheme of residential and retail buildings for the old market. Today there is a new shopping complex comprising 3 acres of glass covered space with craft and antiques stalls, food hall etc. Liverpool Street station is nearby.

Geffrye Museum

Situated near Old Street Station, this compact museum has an excellent collection of furniture and woodcrafts. There is a series of period rooms arranged in chronological order from around 1600 to 1950. The inglenook of the 17th century is a reminder of the household fire place and kitchen nearly 400 years ago (071 739 9893).

Whitechapel Art Gallery

The Gallery with its striking art nouveau facade was opened in 1901 in Whitechapel High Street. It exhibits the spectrum of paintings, sculpture, photography, ceramics and drawings of professional and amateur artists working in studios in East London (071 377 0107).

Museum of Childhood

The museum is dedicated to the concerns of children and childhood. It contains a huge and entertaining collection of toys, dolls and dolls' houses, games and childrens' clothes. Originally built for the Victoria & Albert Museum in South Kensington it was moved to the present site in Cambridge Heath Road in 1868. The extensive collection of dolls' houses are on the ground floor, and upstairs there are toys with special sections for whistles, kazoos and other children's instruments. There are also dolls from other countries. Bethnal Green tube station is nearby (081 980 2415).

St Katharine Haven and Dickens Inn

Tobacco Dock and Wapping Historic Riverside

Wapping Pierhead and Olivers Wharf

Tobacco Dock Shopping Precinct

Tobacco Dock is one of London's shopping precincts formed recently in a Grade I listed building close to Docklands Light Railway (DLR) Shadwell Basin. The Tower of London and St Katherine Docks are nearby. The warehouse was originally built 1811-1814 for the storage of tobacco. It was used to store tobacco, wool and wine and spirit for over 150 years. In the beginning, tobacco was stored on the ground level of the warehouse in large bales known as hogsheads, each 4 feet high and 2.5 feet diameter and stuffed tight. By the 1860's the same floor was used for storage of sheepskins and hides imported from Australia, New Zealand and South Africa.

Hydraulic Pumping Station

The Pumping Station at Pelican Wharf with its original machinery dates from 1893. It was one of five stations providing power not only for the surrounding docks but throughout Central London. Before the adoption of electricity, hydraulic power was the capital's main power system operating everything from dock cranes and lifting bridges to lifts in houses and hotels in Kensington and Mayfair. During the 1930s, more than 33 million gallons of water under pressure raced under the streets of London for the raising and lowering of equipment. As a source of power it was cheap, efficient and easily transmitted along 186 miles of

underground cast-iron piping. DLR Shadwell Station and Wapping tube station are nearby.

Tower Bridge used the pumping station, as did many City offices and West End department stores and theatres. The revolving stages of both the Palladium and Coliseum used hydraulic power as did the fire curtains of Drury Lane and Her Majesty's Theatres. It also operated the organ console lifts of Leicester Square Theatre and the Odeon Marble Arch, the fire hydrants at the National Gallery and the picture lift at the Royal Academy. Much of the pipework is still in existence today and it is used to carry fibre optic and telecommunication cables.

21st Century Orchestra

During the late 1980s there were proposals to build and move the Academy of St Martin-in-the-Fields Orchestra to a high-tech home by building a purpose-built recording and concert hall in the former hydraulic pumping station. No progress has so far been made.

Wapping Heritage Walk

The historic riverside route from St Katherine Docks to Limehouse follows Wapping High Street and Wapping Wall. Starting from Wapping Pierhead, places of interest include the following:

Black Eagle Wharf A new residential development on an old wharf where casks of Truman's beer were handled.

Wapping Pierhead Houses Built 1811-1813 for dock officials and today converted into residential use.

Olivers Wharf The first warehouse conversion into flats in Docklands (1970)).

Orient Wharf A new residential complex on the site of Victorian tea warehouses.

Wapping Police Station stands on the site of the old station built in 1797 to house the first River Police Force in the world.

St John's Wharf A fine conversion of a Victorian wharf with the adjoining Captain Kidd pub.

King Henry's Wharf The Grade 2 listed buildings were owned by Alexander Tug Company.

Gun Wharves and Place A superb conversion of Grade 2 listed Victorian warehouses adjacent to Wapping Station.

New Crane and Metropolitan Wharves Conversion of a number of 19th century tea warehouses into residential and commercial units.

Pelican and Prospect Wharves Two complexes of riverside apartments in Wapping Wall Conservation Area.

Free Trade Wharf attractive Georgian warehouse conversions and new blocks of flats.

For further details please see the book by the author "Discover London Docklands" A to Z Illustrated Guide. Published 1992 and reprinted 1993.

London Docklands Waterside Pubs

(1) The Horniman at Hays, Hays Galleria, Tooley Street, SE1. A new pub built within Hays Galleria shopping arcade with views over Tower Bridge, the City and the museum ship HMS Belfast (071 407 3611).

(2) The Dickens Inn, St Katherine's Way, E1. Recently converted from an old warehouse, the pub overlooks St Katherine Docks marina (071 488 1226).

(3) The Town of Ramsgate, 62 Wapping High Street, E1. Adjacent to Olivers Wharf, this historic pub lies within the Wapping Pierhead Conservation Area (081 488 2685).

(4) The Captain Kidd, 108 Wapping High Street, E1. In a warehouse close to the old Execution Dock, this pub has coveted window seats (071 480 5759).

(5) The Angel, 101 Bermondsey Wall East, Rotherhithe, SE16. This historic pub is said to be the drinking place of Captain Cook (071 237 3608).

(6) The Mayflower,117 Rotherhithe Street, SE16. Named after the ship that took Pilgrim Fathers to America, this historic pub is in St Mary's Church Conservation Area and near Rotherhithe tube station (071 237 4088).

(7) The Prospect of Whitby, 5 Wapping Wall, E1. Probably the oldest on the river, this 16th century pub boasts Pepys, Judge Jeffrey and the painter Turner amongst its former regulars (071 481 1095).

(8) The Barleymow, Narrow Street, E14. Originally a Dockmaster's House, this pub stands at the entrance to Limehouse Basin (071 265 8931).

(9) The Grapes, 76 Narrow Street, E14. Famous for its Dickens' connection, this historic pub has an old river balcony and a fish restaurant (071 987 4396).

(10) Bootys, 92A Narrow Street, E14. Near the Grapes, this pub lies within Narrow Street Conservation Area (071 987 8343).

(11) Scandic Crown Hotel, Nelson Dock, Rotherhithe Street, SE16 (071 231 1001).

(12) The Moby Dick, 6 Russell Place, South Dock, SE16. A new pub overlooking the Greenland Dock (071 231 5482).

(13) The Henry Addington, 26-28 MacKenzie Walk, Canary Wharf, E14. An American style bar overlooking the historic West India Export Dock with its occasional boats and barges (071 512 9002).

(14) The Cat & The Canary, Fisherman's Walk, Canary Wharf, E14 Recently opened, this pub looks over the West India Import Dock towards the magnificent historic sugar warehouses of future Port East Complex (071 512 9187).

(15) Drummonds, Heron Quays, Marsh Wall, E14. A large bar with a terrace overlooking the West India South Dock (071 538 3357).

(16) The Waterfront, South Quay Plaza, 187 Marsh Wall, E14. A modern bar which overlooks the West India South Dock with a fine view of Canary Wharf (071 537 2823).

(17) The Spinnaker, Harbour Island, Harbour Exchange Square, E14. Part of a new development, this pub overlooks Millwall Dock (071 538 9329).

(18) Manzis, Turnberry Quay, E14. A branch of the Leicester Square fish restaurant, this bar has fine views of Millwall Dock (071 538 9615).

(19) The Gun, Coldharbour, E14. Steeped in history, this pub is said to be the tavern used by Lord Nelson for meetings with Lady Hamilton (071 987 1692).

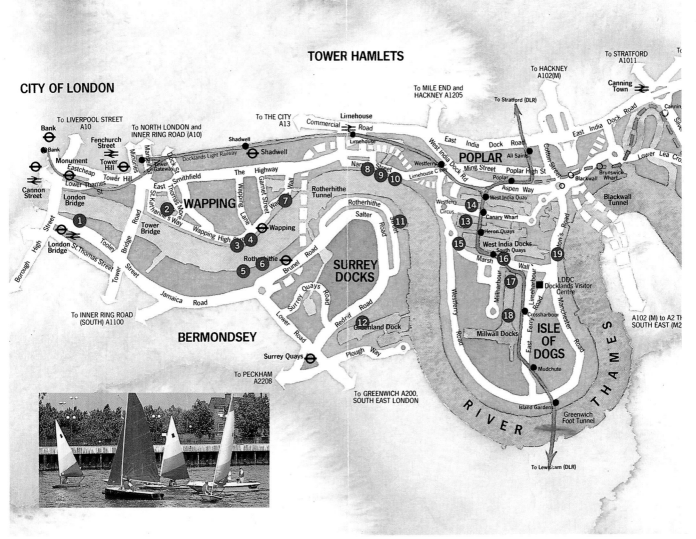

Map showing Docklands Waterside Public Houses

London's Canary Wharf

Canary Wharf Tower

Old Canary Wharf

On the Isle of Dogs you encounter the new developments in the former West India Docks. In the 19th century, Canary Wharf was called the Rum Quay and was the home of the mahogany sheds and Rum Warehouses. Aromatic cargo from West Indian ships was stored in the warehouses and vaults and various coopering operations were carried out on behalf of the merchants concerned. The vaults were well built but not as extensive as those of London Docks. In 1936 the quayside caught fire and it was also heavily bombed during the war. Subsequently sheds Nos. 10 and 11 were built. The Canary Wharf became a tenant berth for the Fred Olsen Line discharging Canary Islands' produce such as tomatoes and bananas. The wharf had a distinctive smell of rotting tomatoes on the quayside! The fruit traffic from the east and west Mediterranean ports was quite heavy. After the closure of the docks in 1970 shed No.10 was leased to Limehouse Studios for film making, but in 1985 it was demolished to make way for the new Canary Wharf complex, one of the financial centres of London. New Year laser display provided a spectacular start for 1992.

New Canary Wharf

The most common reaction among visitors to Canary Wharf is amazement at the sheer scale of the development. A short walk around the site shows the beautiful buildings, tree-lined streets, the squares and the sky, the river and the docks, all forming an urban centre but retaining a sense of pleasure and relaxation.

Cabot Square and Hall

The tree-lined West India Avenue runs from Westferry Circus eastward to Cabot Square. This Square is a popular meeting and resting place for visitors and people working there. Its central area is paved with polished and flamed Toricicoda granite from Brazil, and with Pearl Anglais granite from Sweden. At the centre of the square there is a 120-jet illuminated granite fountain. Around the fountain there are sculptures incorporating the air vents from the car park below. This open space, the same size as St. James's Square is adjacent to Cabot Hall to the east. The Hall, which accommodates receptions, concerts and other public events, also contains four floors of shops and the Canary Wharf Station of the Docklands Light Railway. Cabot Place on

the east side of the Hall is crowned by the glazed dome of the fine rotunda.

Canary Wharf Station

Set in the heart of the development this Dockland Light Railway Station gives access to the retail building of Cabot Place, to Cabot Hall and the Tower. The station is covered with a high arched and glazed roof - a style reminiscent of the great Victorian railway stations. Beneath, three tracks and six platforms provide capacity for up to 40 trains every hour.

No. 10 Cabot Square

This building on the north side of Cabot Square has a fine façade and interior. The exterior cladding consists of London Brick, hard-set in fine reconstituted limestone which is decorated with Celtic motifs and green polished granite plaques. These have been set into the circular recesses between the arches of the arcades. The magnificent lobby is two storeys high and is floored with no less than seven different types of Italian marble. English oak, cherry wood and teak are used for doors throughout the building. Italian marble has been used for finishes.

No. 20 Cabot Square

The northern frontage of this 10-storey building is at the south east corner of Cabot Square. It has two main entrances to the offices at No.20 Cabot Square and No.10, The South Colonnade. The arcade at ground level caters for a variety of retail shops. The curved facade on the south side extends across the dockside with views of the river and Greenwich. The building is clad in highly-figured honey colour Vermont marble.

No.25 The North Colonnade

North of the Tower is a modern office building with a special construction layout; its 15 floors are built out from a central core which carries the lifts and all services. The building is finished in glass and grey granite panels which were imported from Italy. Across the waters of the old West India Import Dock lies Port East, the proposed development of the two historic warehouses 1 and 2. A footbridge connects the block to the warehouses.

No.30 The South Colonnade

At the south eastern corner of the development is this building which is separated from No.20 Cabot Square by the Docklands Light Railway. The two 10-storey buildings are said to have a nautical style, a reminder of the historical past of the site, with the great mercantile buildings which once lined the docks and wharves of London. Both buildings are clad in highly-figured grey Vermont marble.

Canary Wharf Tower

This 50-storey Tower at No.1 Canada Square, rises to a height of 800 feet (246 metres) - Britain's tallest building, soaring 200 feet higher than the National Westminster Tower in the City. It is a new landmark on London's skyline. The crowning pyramid, floodlit at night, is clearly visible for many miles around. Nine different varieties of fine Italian marble from Forte de Marni, were used for the floor and walls of the 3-storey Tower lobby. They include honeyed Nero Marquinia, Rosso Rubino, Fior di Pesco and polished Verde Quetzel, all creating an ambiance of colour and elegance. Thirty-two high speed lifts, running in staggered sequence within four separate internal cores, carry passengers up and down the building. The top floor can be reached in just 45 seconds. The 46 office floors within the Tower are equipped with the latest air conditioning and have the most advanced data and communications systems. The views from the Tower are breathtaking. Looking south from the top floor you can see across the Isle of Dogs to Greenwich and further beyond South London to the countryside. London appears to spread at your feet!

Westferry Circus

Westferry Circus has English paving stones with Spanish granite kerbs. The central area is lawned surrounded by seats and sculptures. Set in a ring around the centre tall trees have been planted in self irrigating containers.

Britannia International Hotel

Located a short distance from Westferry Circus this 435 bedroomed hotel has three restaurants and a night club. Facilities include meeting rooms for up to 800 people.

Water Sport

The water areas of the West India and Millwall Docks are used by clubs for sailing, Water skiing, Jet skiing, wind sailing, power boating, and canoeing. Some fishing is also permitted. Please contact the Docklands Development Corporation (071 512 3000).

Canary Wharf

Map of Canary Wharf

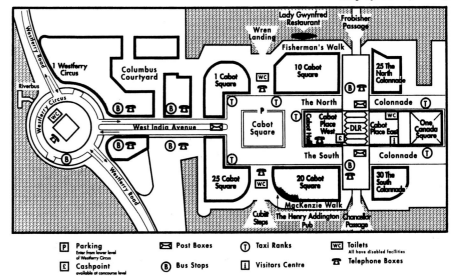

Dockside Shipping Scenes 1950s and 1960s

West India and Millwall Docks

About 5 miles downstream from Tower Bridge on the north side of the river, lies the Isle of Dogs, encompassed on three sides by the great sweep of Limehouse, Greenwich and Blackwall Reaches. Here were West India, East India and Millwall Docks. The water area of these docks was 151 acres, with 48 berths in a total length of quay of 8 miles. They could take large vessels up to 550ft long and 28½ ft draught. Imports and exports exceeded three quarter million tons each.

Approximately 50 shipping lines regularly used these docks. They traded with North and South America, East, West and South Africa, India, the Mediterranean, Persian Gulf, the Middle and Far East, France, Spain, Portugal, Scandinavia and of course the original connection with the West Indies. Sugar was handled in bulk on the north quay by self-dumping grabs attached to 5-ton cranes. An output of up to 500 tons per gang per 8-hour shift was achieved in the later 1950s by mechanisation. Green fruit was handled at specially constructed reinforced concrete and brick sheds. The Millwall Docks' skyline was dominated by the dockside flour mills of McDougalls and the Port of London Authority 10-storey Central Granary where up to 24,000 tons of grain could be held in store. Along the Granary four fixed elevators discharged grain from the ships.

Millwall Inner Dock 1950s

The picture at the top of the page shows shipping scenes in the 1950s alongside the sheds adjacent to the Central Granary. The building was demolished early 1980s and replaced with present day developments including Glengall Cross on the west side of the dock. The Norwegian vessel *K C Rogenes* from the Port of Hangesund is being loaded with general cargo for George Town, Trinidad. The hydraulic crane is loading from the sets outside the sheds. "Dirty cargo" of drumwork (40 gallons) received ex vehicle awaiting loading at No.2, the main hold on the ship. Drum crates of chemicals on quay and carboys are also awaiting loading. The stevedores put them in sets and then loaded them on the 'tween deck (above the lower hold). The Port of London Shunters (the foreman uniformed) are bringing a box truck laden with cargo for loading by the crane.

Information on London Docklands

For further information on the history and regeneration of London Docklands, the reader is recommended to see the books by the author, 'Dockland', 'London Docklands' and 'Discover London Docklands'. Details are given on page 2 of this publication.

Shipping scene in the 1950s

Map of the West India and Millwall Docks, circa 1929

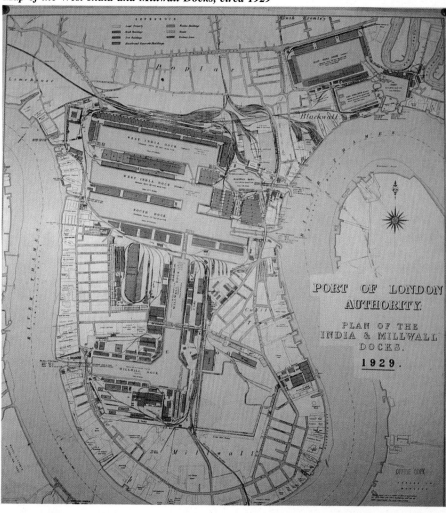

Conservation & Heritage Trail on Isle of Dogs

Using Docklands Light Railway, a heritage trail can be mapped which passes historic buildings and landscapes on the north, south and east sides of the Isle. The few remaining historic buildings and landmarks in the conservation areas have been safeguarded by listing as the only tangible link with the island's past and are briefly described below.

West India Conservation Area

The West India Docks (DLR North Quay) were the first commercial wet docks built in London at the beginning of the 19th century. They pioneered a new concept of enclosed docks, with monumental multi-storey warehouses on the quays, within the security of a moat, boundary wall and a fence, all guarded by the Dock Constables. At the north west corner of the Isle of Dogs there survives a unique collection of the old features of an enclosed dock with its ledger building, the round guard house, police constable cottages (1819), Dockmaster House and excise offices (1807), Engineers' quadrangle building including workshops and stores (1824), and two magnificent warehouses, 1 and 2, (1802). The Georgian warehouses are the only surviving multi-storey brick buildings of their period.

The two northern import and export basins of the West India Docks were completed in 1802 and 1806 respectively. The Millwall inner and outer basins were built comparatively later and completed in 1868 for bulky commodities such as grain and timber.

Island Gardens Conservation Area

These gardens (DLR Island Gardens) were developed in the 19th century and are among the finest in the Borough of Tower Hamlets. The Conservation Area contains five statutory listed buildings; Ferry House Pub (early 1880s), Christ Church (1854), Waterman's Arms Public House (1850s), Newcastle Draw Docks and the Foot Tunnel entrance (1902). You should pause to admire historic Greenwich with its magnificent waterfront Royal College and Cutty Sark before plunging into the Foot Tunnel to cross the River Thames to Greenwich.

Coldharbour Conservation Area

This area near the Blue Bridge along Preston Road, stretches between two historic entrance locks for ships in the former docks in the Isle of Dogs. It includes the Gun Public House (c19th façade on older building), Nelson House (early 1860s) and the first River Police Station (1894) form part of the fine river front. The Bridge House (1819) is on the other side of Preston Road.

Map showing recent developments on the Isle of Dogs

South Quay Plaza & Harbour Exchange

Heron Quays
Heron Quays was the first commercial development on the Isle of Dogs and has a waterside business and residential community. The low rise buildings are surrounded on three sides by the waters of the old Export Dock and West India Docks.

Britannia International
Located a short distance from Westferry Circus at Arrowhead Quay, Marsh Wall, the hotel has 435 bedrooms, three restaurants and a night club. Other facilities include meeting rooms which can accommodate up to 800 people and other suites overlooking the West India Docks and Canary Wharf.

Scandinavian Centre
A striking Swedish-style office building, standing entirely on piles in the waters of West India South Dock between Heron Quays and South Quay. The 4-storey building has a central naturally lit atrium for display and recreational purposes.

Cascades
The first luxury residential tower to have been built in Docklands is the Cascade. It is a twenty-storey block with fine views of the City, Greenwich and the Thames, with two penthouses on the top floor riverside.

Waterside
Located on the South Quay of the old West India South Dock between the Scandinavian Centre and South Quay Plaza, this complex consists of Quay House, Ensign House and Beaufort Court comprising of 48 business apartments. Private balconies and mooring facilities with quayside walkway are attractive features of the development.

South Quay Plaza
Along Marsh wall, the South Quay Plaza is one of the largest office schemes with waterside settings and its own Docklands Light Railway station of South Quay. There are three self-contained office blocks of seven, ten and thirteen floors. The development includes a shopping plaza inside the tallest block with a public house and restaurant facilities on the southern part of the site.

Thames Quay
Thames Quay is an office building east of South Quay Plaza occupied by the London Docklands Development Corporation. Sitting on the corner of South Way and Millwall Cut it is a bright, highly visible building with a design which is reminiscent of the form and detailing of an ocean-going ship.

Meridian Gate
Meridian Gate is a mixed commercial complex located on the waterfront of South Quay, adjacent to Thames Quay. There is a plaza surrounded by shops and a restaurant.

Harbour Exchange
One of the largest office developments recently completed in London Docklands. Harbour Exchange has eight buildings of different sizes. The central piazza, surrounded by shops, a restaurant and a pub, opens out onto a promenade fronting the old Millwall Inner Docks with its own water vistas.

London Arena
A large concert, exhibition and sports hall where classical and popular music concerts are held.

London Docklands Visitor Centre
The Centre in Limeharbour has a comprehensive audio-visual presentation about Docklands which is available for pre-booked groups in Japanese, French, German and Swedish as well as English. A collection of relics from the old historical docks gives an insight into the history of the area over the past 200 years.

Glengall Cross
The Glengall Cross is a mixed commercial, retail and residential development and spans the waters of Millwall Inner Dock with a new lifting steel bridge.

The Brunel Centre
A scheme which combines office accommodation, hotel, restaurant, and medical facilities with an adjacent leisure and watersports centre. The 237 bedroomed hotel is set in an attractive position overlooking the south-eastern end of Millwall Docks.

Greenwich View
This is a development of offices and high-tech business suites in a corner position fronting Millwall Outer Dock with the Daily Telegraph Printing Works to the west and Glengall Cross Development to the north. A spectacular island office building is constructed in the water of the dock.

Docklands Light Railway South Quay Station

Financial Times, Abbey Mills and Royal Docks

Financial Times Print Works
The impressive printing works has a 96 by 16 metre continuous glass main central façade exposing the giant printing processes to public view along the A13 trunk road. The effect is particularly spectacular at night when the presses are busiest. Completed in 1988, the building is dominated by the long press hall on the north side. This is flanked at one end by the paper store and at the other by the despatch area. Aluminium sheeting is used to clad these parts. To the south, there are three storeys of office and ancillary areas, with another glass wall.

East India Office Complex
This development in the former East India Docks has an impressive range of office buildings with an underground car park. The main buildings have white marble with bay windows and stone medallions. The complex adjoins the Financial Building to the north.

London's Lighthouse
The only lighthouse in London is located in an historic riverside site at Blackwall. It provides a link between the ancient Corporation of Trinity House and Professor Michael Faraday, the founding father of electricity. Built around 1860, the octagonal lighthouse is a centrepiece of Trinity Buoy Wharf at the confluence of the River Lee with the River Thames. The lighthouse had no navigational function but was used to train lighthouse keepers in the art of maintaining lanterns.

Abbey Mills Pumping Station
A landmark in East London, the 1856-68 pumping station was designed and built by Sir John Bazalgette to raise the level of the northern outfall sewer as part of the London sewerage system. The exterior is a mixture of Byzantine and Gothic designs with a mansard roof. The large galleried pumping hall originally accommodated beam engines which were replaced by electric motors in 1933. It is near West Ham tube station in Abbey Lane, E15.

Passmore Edwards Museum
The museum currently houses the collection of the Essex Field Club. Permanent excellent displays include natural history, geology, archaeology and East London history. Temporary exhibitions are also held. Located near Stratford underground station, it is adjacent to the Stratford Campus of the University of East London in Romford Road, E15 (081 519 4296).

Three Mills
These mills are the largest tidal mill complex

Shipping scene in the Royal Docks, 1960s

in England. Only two mills remain, the third, a windmill, has been demolished. Standing by the River Lea, this historic landmark in Newham has had trading links for nearly eight centuries. The Clock Mill, with its fine original clock tower, was built 1773 and is 80ft long and four storeys high. It was rebuilt in 1817 in yellow London stock bricks and contains the tide mill water wheels and two tall drying kilns. The building has been converted into offices.

The House Mill, built 1776, has much of its machinery and water wheels intact and is being developed as a working museum. Located in Three Mills Lane, E15, further information may be obtained from the Passmore Edwards Museum (081 519 4296).

Royal Docks
At the eastern end of Docklands are three massive docks - The Royal Docks. They date back to the 1850s with the last one being opened by King George V in 1921. For over a century they were the centre of world trade and at the hub of the British Empire. They contain London City Airport and a huge water environment. There have been many proposals to develop the area into a water city of the 21st century with homes, shops

and leisure facilities. For further information please contact the London Docklands Development Corporation (071 512 3000).

Tate & Lyle
Tate & Company started sugar refining at the Thames Refinery, Factory Road in 1878 and Lyle & Sons established the same three years later at nearby Plaistow Wharf. In 1921 the two firms amalgamated and their present building, opened in the 1930s, was built in Portland Stone. Sir Henry Tate donated £80,000 for the building of the Tate Gallery. Abram Lyle invented Golden Syrup in 1883 and his familiar emblem is carved high on the north western façade of the building.

North Woolwich Railway Museum
The museum in Pier Road, E16, tells the story of the Great Eastern Railway from its origins in 1839 and you can see a locomotive steaming along the museum's track.

Museum in Docklands
Housed in the historic W Warehouse on the north quay of the Royal Victoria Dock, the museum contains a collection of old dock tools and machinery. A guided tour of the museum may be arranged by appointment (071 512 3000).

Barking Abbey and Fords of Dagenham

Barking Abbey and St Margaret's

Founded in 666 by St Erkenwald, the Abbey was occupied by William the Conqueror whilst waiting for the Tower of London to be built. Its demise was in 1539 when Henry VIII suppressed the Roman church. The Crown controlled the area until it was sold to Sir Francis Fanshawe in 1628. The Fanshawe family played an important role in the development of the present Borough of Barking and Dagenham and their relics can be seen in the local Valence House Museum. Several walls of the ancient Abbey church were excavated in the 1960s and can be seen today. Also remaining is the Curfew Tower which stands at the entrance to St Margaret's Parish Church. This church was built in the 13th century within the Abbey. The well-known explorer, Captain James Cook, was married to Elizabeth Batts in the church in 1762. The church is within a few minutes' walk of Barking tube station and is open to the public by appointment with the Rector (081 594 2932).

University of East London

Nearby the Abbey and situated in Longbridge Road is the Barking Campus of the University of East London which is one of the largest universities in Britain and its

origin dates back to 1902. Visitors are welcome by appointment (081 590 7722).

Valence House Museum

This timber framed manor house, dating from the 17th Century houses Barking and Dagenham's local history museum and art gallery. It is located in Becontree Avenue, Dagenham, Essex, within a short walk of BR Chadwell Heath Station (081 592 4500).

Barking Barrier

The present massive barrier at the mouth of Barking Creek forms part of the London Flood Defence System and complements the Thames Barrier. There is a proposal to build a second barrier across the River Roding to create a leisure and watersports centre on the East London Waterway. A heritage centre is also being proposed to explore the town's historic links with London's fishing industry, Napoleonic Wars and Henry VIII.

Ford's Motor Works

Ford is a major car manufacturer in Britain and its Dagenham plant on the north bank of the Thames is the hub of nationwide activities by the company. It is the largest employer of labour in the Dagenham area. In May 1929, Henry Ford's son Edsel visited

London and inaugurated the site of the huge new factory which was built on reclaimed marshland. In October 1931 their first vehicle, a Model A truck, was driven off the production line. Approximately a year later, Ford created the first small car, Ford Eight Model Y, which sold at a price of £120. Post 1950s Ford produced a new range of cars including the Consul, the Zodiac, the Anglia, the Prefect and the Popular. Later, the successful Cortina lasted for 20 years. Currently, the Dagenham plant has developed its activities to cover other parts of Europe. It ships around 150 trailers and 50 containers each day in each direction to Zeebrugge. The company has three 'roll-on roll-off' vessels and two container ships on charter. Dagenham is one of the most advanced robotic motor factories in the United Kingdom. Industrial visits may be arranged for tourists and other groups through their Public Relations Agency (0277 253000).

Upminster Windmill

Built 1802-3 by James Noakes, it is a landmark in Upminster. From 1818, it contained one of the earliest steam engines used in Essex mills. The mill has good views across Ingrebourne Valley to St. Andrew's Church. Upminster is the nearest tube station.

Fords of Dagenham

SOUTH EAST LONDON

Royal Greenwich

Butlers Wharf, Design, Tea & Coffee Museums

Bermondsey and Shad Thames
Adjacent to Tower Bridge is Shad Thames, where you can wander through the cobbled canyons of Victorian warehouses with their bridges, now magnificently renovated as riverside homes, restaurants and shops.

Butlers Wharf
Butlers Wharf is a colourful tourist attraction on the south bank of the River Thames, adjacent to Tower Bridge and with its own Riverbus Pier. The area is rich in history and, in the 19th century, was a busy trading wharf for the import of tea, coffees and spices. Today, the warehouses have been imaginatively restored and converted into commercial and residential buildings with cobbled streets and beautiful riverside frontages including a winebar and restaurant.

Design Museum
Adjacent to Butlers Wharf, the museum is the first in the world to examine the role of design in our daily lives and offers visitors and schoolchildren a fascinating insight into design, technology and consumer culture. It provides interesting resources for the specialist and members of the general public.

An important feature of the museum is the study collection. This analyses design in mass production in a variety of contexts including social, economic, functional, commercial and technological. It is arranged in broad thematic parts such as chairs, home, office and transport. There is also a gallery for temporary exhibitions looking at the past, present and future of design.

The museum offers special facilities for schools and colleges. It runs an educational service with free admission for pre-booked school and college parties and academic staff. It has a lecture theatre equipped with audio visual equipment. For the researcher there is a reference library of books, journals and video tapes. From Tower Bridge underground station visitors can walk over Tower Bridge and the museum is adjacent to Butlers Wharf. The riverside cafe and bar have excellent views of the City (071 407 6261).

Tea and Coffee Museum
Next to the Design Museum, this new museum is particularly interesting for lovers of antiques. It houses the Bramah Collection of pictures, silver and ceramics and shows the equipment for making and serving tea and coffee. Teapots, teaware and coffee-making equipment are displayed. A collection of 200 prints relating to tea between 1740 and 1840 are on show and, together with paintings and photographs, highlight the cultural and social significance of the subject. A coffee bar and tea room are there to refresh the visitors. The museum shop is stocked with choicest teas and coffees. The museum is at the Clove Building, Maguire Street, Butlers Wharf, SE1 (071 378 0222).

The Circle
This is a mixed development of housing, shops, offices and workshops, built mainly of yellow stock brickwork with deep set windows following the local 19th century warehouse construction. The impact of the central curved walls of blue glazed bricks rising seven storeys up, is highly impressive. The geometrical shape of the buildings with a larger than life cart horse sculpture at its centre gives the development its name. Stock brick facades forming the bulk of the elevation have balconies supported by diagonal struts reminiscent of round timbers found in old warehouses.

Anchor Brewhouse and Butlers Wharf

Dulwich Village, College and Picture Gallery

Dulwich College

South London Boroughs
South London consists of a number of one-time villages, now London Boroughs, which developed and expanded in the 18th and 19th centuries. These areas have numerous historic sites and beautiful parks intermingled with modern buildings.

Dulwich Village and College
Dulwich Village is a pleasant suburb of London with some well preserved Georgian houses and cottages, particularly along College Road to the mill pond opposite Dulwich College. Founded in 1619, the College buildings, in Italian Renaissance style, are mainly mid 19th century by Charles Barry. West Dulwich station is nearby.

Picture Gallery and Mausoleum
Built 1811-14 by John Soane for Dulwich College this was one of London's earliest public art galleries. The gallery contains a collection of paintings belonging to the art dealer Noel Desenfans which was intended for the Empress of Russia but was later left to the College. The building has a small mausoleum and almshouses. The gallery is located close to West Dulwich Station.

Dulwich Park Walk
Dulwich is a very select, hilly village with steep roads and the only remaining toll gate in London. The formal park is most delightful and the woods at Tree Hill are thick with trees. The grounds at Belair near West Dulwich Station have become a public recreation space and contain the source of the River Effra. There is a car park at Belair and parking is allowed on the main drives in Dulwich Park, Entering the park, after a short walk you find yourself along the lakeside. Further along a woodland is reached were you see the trackbed of the old Crystal Palace branch railway line by the side of the fine Horniman Gardens. Further north are One Tree Hill Park and Peckham Rye Common. West Dulwich is reached from Victoria.

Severndroog Castle
In a commanding position off Shooters Hill Road, SE18, the triangular Gothic tower was built 1784 to commemorate the capture of Severndroog Castle, Malabar, India, in 1755. Set in the parkland location of Castlewood, the tower has superb views of London and Woolwich but is not open to the public. Eltham Park Station is close by.

Beckenham Place Park
This park, within the London Borough of Lewisham, has over 200 acres of grassland and woods similar to Richmond Park and serves as an excellent walkers tour. There is also the River Ravensbourne and if you want to see more of this river, you can visit the nearby Ladywell Park. BR Beckenham Hill Station is easily reached from Holborn Viaduct. There is a car park near the Mansion and formal gardens.

Horniman Museum
In London Road, Forest Hill, SE23, this museum has extensive ethnographic, natural history and musical instrument collections. Exhibitions celebrate aspects of cultures from Africa, the Americas, Ancient Egypt, the Far East and the Pacific. There are also displays of Folk, Oriental and European musical instruments. Built 1896-1901 to the design of C H Townsend for John Horniman, the building has a large clock tower with a circular cornice. The central mosaic panel on the front elevation was painted by R A Bell. The lecture hall and library were added later in 1910. BR Forest Hill Station is near the museum (081 699 2339).

Crystal Palace and The Old Palace, Croydon

Crystal Palace, early 1930s

Crystal Palace Park

The Crystal Palace, designed for the Great Exhibition of 1851, was built of glass and iron in Hyde Park. It was afterwards taken down and re-erected in this Park in 1852-4 on the site of Penge Place. Sadly the building was burnt down in 1936, but the two terraces of the formal garden, designed by Sir Joseph Paxton at the time of re-erection, have survived with remains of steps, balustrades, fountains and statues. The site is now a large National Sports Centre, Crystal Palace Stadium, where various national sporting events take place.

BBC Tower

One of Britain's best known steel towers is the BBC's 216m high Crystal Palace tower which transmits television throughout Greater London.

The Priory Museum

Dating in part from 1290, the Priory is the home of Bromley Local History Museum. The building was used as a rectory after the Reformation and later became a private house. The displays include the Avebury collection and local archaeological material. Located in Church Hill, Orpington, with a British Rail station close by (0689 31551).

Croydon

This area in South London has developed over the last few decades with many new buildings accommodating large companies and government offices that have moved from the City and West End areas of London.

The Old Palace Croydon

The former residence of the Archbishops of Canterbury, its foundations date back over 1000 years. Used by the Archbishop as a country residence and as his headquarters when engaged in diocesan business. The building can boast associations with Archbishop Laud during the 17th century. The main historic buildings are the 15th century Great Hall, the Guard Room, the Tudor Long Gallery and the domestic Chapel. The Palace is now the home of the old Palace School for Girls and the west wing contains some of the earliest medieval brickwork in England. British Rail West Croydon Station is nearby.

Whitehall of Cheam

Built around 1500 on land attached to East Cheam Manor, Whitehall is an example of one of the earliest types of houses to have two storeys throughout. The museum displays material from local archaeological excavations including Saxon relics. BR Cheam Station is near the house in Malden Road, Sutton (081 643 1236).

Croydon, South London

Rotherhithe, Greenland Dock and Marina

Rotherhithe and Surrey Docks

The Surrey Commercial Docks were spread over some 380 acres of Rotherhithe. with eleven docks all interconnected by water cuttings, they had two lock entrances into the Thames; one, the Surrey Entrance in the Lower Pool of the Thames for smaller ships and craft, and the other, the Greenland, in Limehouse Reach for larger vessels. The latter entrance lead into the Greenland Dock, which, with South Docks, dealt with general cargo. The remaining nine docks of the system concentrated on the discharge and storage of timber. Regular steamship lines used these docks to trade to and from Canada, North and South America, India, the Baltic States and Russia, the near continental ports and the far Pacific Islands. Timber from Scandinavia, Russia and Canada was carried mostly in complete shipments by specially chartered tramp steamers. Since 1981, the whole area has been at the centre of an amazing transformation by the London Docklands Development Corporation, with the building of housing estates of many thousands homes. There is a thriving community with green open spaces, tree-lined walks and nature reserves which combine readily with centuries' old churches and riverside public houses.

South Dock Marina

Greenland Dock and the adjoining South Dock Marina, areas have been set aside for sailing, rowing and canoeing with 19 acres of water available for use by clubs and for professional instruction. The Marina, four miles downstream of Tower Bridge, can accommodate craft of up to 100ft long, 20ft beam and 11ft draught and is operated 24 hours a day. The lock from the Thames operates for three hours either side of high water and moorings are linked to the shore by pontoons.

Surrey Quays

The shopping precinct is located in landscaped surroundings with nearby watersports facilities. Alongside Tesco superstore, shoppers will find British Home Stores (BHS), Boots the Chemists, Top Shop, W H Smith, restaurants and other shops. Surrey Quays tube station is within a short walking distance.

London Glassblowing Workshop

At 108 Rotherhithe Street, SE16, near Rotherhithe tube station, you can watch glass being blown and then cross to the gallery in the old Granary building where it is on sale. The last weekend in November and first in December are the special times to pick up bargains when sales of 'seconds' and near perfect items are held (071 237 0394).

Rotherhithe and Surrey Docks today

Surrey Commercial Docks, circa 1906

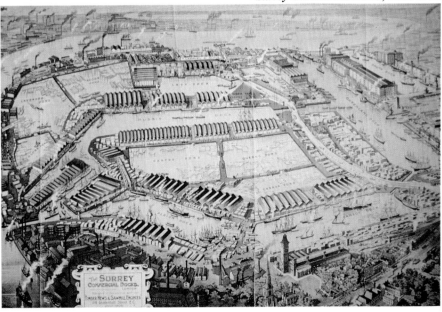

Greenwich Museums and Royal Observatory

Royal Greenwich

In Tudor and Stuart times, the Royal Palace at Greenwich was the seat of the monarchy, affecting the future development of British trade and empire. Its strategic location at the bend of the River Thames dominated the approach from both land and water, and overlooked the whole of what is today known as London Docklands on both banks of the river. Greenwich was the birthplace of Henry VIII, Edward VI and Elizabeth I. The area is steeped in naval history. Cabot and Columbus came to petition King Henry VII for support for their voyages to the New World of the Americas. Sir Walter Raleigh, Captain Bligh and Captain Cook lived here. Sir Francis Drake was knighted by Queen Elizabeth I on board his ship, the "Golden Hind". Peter the Great of Russia learned about navigation at the Royal College, Lord Nelson's state funeral procession went along the loop of the river around here. Cleopatra's Needle travelled the same route on its last few miles of travel from the Port of Alexandria in Egypt, to its present location on the Victoria Embankment in the City.

Greenwich is rich in heritage and full of interest and entertainment. Here you will find the National Maritime Museum, the Royal Naval College, the great Royal Observatory, the famous tea clipper, The Cutty Sark, and Gipsy Moth IV in dry dock. The name of Greenwich is possibly derived from the Anglo Saxon Green Village or Green Reach. It is said that Viking ships layed off here in 1011-14. It has a range of craft shops as well as the Greenwich Market which is open at weekends and is a place for browsing through books and antique jewellery. The riverbus, which can be caught at Greenwich Pier, is an ideal place to start and end your visit. From the riverside promenade you can watch the passing of barges, tugs and pleasure boats. The river bus starts from Westminster, Charing Cross and Tower Pier. You may also cross the River by the Foot Tunnel at Island Gardens Station of the Docklands Light Railway. The walk through the tunnel takes about 6 minutes. Greenwich British Rail Station is also convenient for travel from Waterloo.

Cutty Sark

This 19th century sailing clipper has been standing in dry dock since 1954. Built in 1869, the tea clipper has beautiful tall masts and graceful lines. The ship made the fastest voyage for a sailing ship between Australia and England in 1887. Now she is a popular attraction for tourists and visitors. Her greatest days were as a wool clipper bringing the new season's clip from Sydney to London in record time. You can wander on the upper decks and hold the wheel and explore below decks, the cabins are on view and there is an audio visual presentation of the ship's history. On the lower decks a collection of merchant ships' figureheads are on display. Guided tours can be arranged.

Gipsy Moth IV

This vessel came to Greenwich in 1968 after being sailed single handed around the world in 226 days by Sir Francis Chichester.

Royal Naval College

A fine and interesting group of historical buildings of 17th and 18th century construction started by Webb, 1664, and developed by Wren, 1692, Vanbrugh, 1728 and the Chapel by Stuart, 1789. Once this was a palace for the Tudor Sovereigns and the birth place of King Henry VIII. During the 18th century it was adapted as the Greenwich Hospital for Naval ratings. See the magnificent Painted Hall by Thornhall where the body of Lord Nelson lay in state in 1805 before the procession along the river to St Paul's Cathedral.

National Maritime Museum

This great museum has the finest maritime collection in the world. An extensive range of model ships, navigational instruments, naval weapons, uniforms and paintings are exhibited. See how life has changed at sea during the past 500 years, since the days of Drake, Raleigh and Nelson. The museum incorporates the Queen's House and the old Royal Observatory.

The Queen's House

Built by Inigo Jones, this beautiful house was completed for the Queen of Denmark in 1619. The spectacular circular Tulip staircase is similar to the one designed by Palladio in Venice.

Old Royal Observatory

This was the first Royal Observatory and the home of Greenwich Mean Time where you can stand astride the Greenwich Meridian. Designed by Christopher Wren, it was built for King Charles II in 1658. It is part of the National Maritime Museum.

Flamsteed House

The oldest part of the Old Royal Observatory buildings is Flamsteed House which was the residence of the Astronomer Royal until 1948. The Octagon room designed by Wren has been restored as it would have been in the 17th and 18th century with contemporary clocks and telescopes.

National Maritime Museum

Blackheath, St Alfege Church & Vanbrugh Castle

St Alfege Church

The 17th century masterpiece by Hawksmoor, is dedicated to St Alfege. Following their raid on Canterbury in 1012, the Danes brought St Alfege to Greenwich and murdered him on the site where the church stands. The church was restored following the bombing of World War II.

Charlton House

A fine Jacobean manor house built 1612 in red brick along Charlton Road. The H-plan house has beautiful ceilings, staircase and a large number of chimneys. There are many ornaments. The house is today used by the London Borough of Greenwich as a community centre.

Vanbrugh Castle

Overlooking the Royal Hospital the Gothic castle was built 1717-26 by Sir John Vanbrugh for his own residence, he was appointed as Surveyor to the hospital as a successor to Wren in 1716. The private house has steep proportions, narrow windows and round towers, and has recently been converted into luxury apartments.

Trafalgar Tavern

This riverside tavern to the east side of the Royal College was built in the 1830s and was a place where writers gathered, including Charles Dickens who mentions the tavern in 'Our Mutual Friend'. It was once famous for its whitebait suppers. Members of Parliament would charter boats to come here and enjoy themselves after a sitting!

Trinity Hospital Almshouses

The fine riverside almshouses and chapel of Trinity Hospital were founded by the Earl of Northampton, Henry Howard in 1631. Today it looks after 20 pensioners.

Greenwich Royal Park Walk

This park has breathtaking views across London to the hills of Hampstead Heath. Once part of the royal forest stretching from Epping to Eltham, the park was enclosed by James I early in the 17th century. The deer park, the formal gardens and lake were created by Charles II to the design of Le Notre, landscape gardener to the French King Louis XIV. Running through the park and its Royal Observatory is the zero Meridian Line. On the west side of the park is the Rangers House with its spectacular rose gardens.

Blackheath

Leave Greenwich Park at Blackheath Gate and make a walking tour of Blackheath, the site of the uprisings by Wat Tyler and of Jack Cade, and the meeting place of Henry VIII with Anne of Cleves! During the 11th and 12th centuries the Danes raided London, anchoring their ships at Greenwich and encamping on Blackheath using it as their base for frequent raids into Kent. During the 13th century Wat Tyler gathered men here to make up the army of his Peasant's Rebellion. Blackheath saw the triumphant return of Henry V after the battle of Agincourt.

Royal Blackheath Golf Club

Blackheath is the location of the world's first golf club, the Royal Blackheath. It was founded in 1608 under the reign of James I, who was also James VI of Scotland and an enthusiastic golfer. He introduced the sport to England.

Queen Elizabeth I

Trafalgar Tavern interior

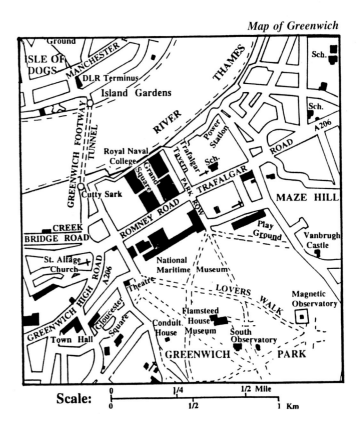

Map of Greenwich

Elizabethan Seafarers, Pirates and Buccaneers

Captain Kidd's body was hung at Tilbury

Elizabethan Seafarers

The great seafarers and adventurers of Elizabeth I's reign were Raleigh, Drake, Hawkins, Frobisher, Grenville and Gilbert. They were navigators and fighters who explored the New World, circumnavigated the earth and pillaged the Spanish treasure ships. In 1587 Sir Francis Drake sank 10,000 tonnes of shipping in a raid on Cadiz, thus "singeing the King of Spain's beard" and delaying the sailing of the Great Armada by a year. A similar raid on Cadiz was made by Sir Walter Raleigh in 1596.

Old Pirates

Recently, the National Maritime Museum held an excellent exhibition based on the lives and times of pirates and buccaneers in the old days. The name pirate comes from the Greek 'peiran' meaning to assault. In the Middle Ages the English Channel had many British and French pirates - who preyed on ships for their cargoes of fish, wines and wool.

Pirates have often been portrayed as a glamorous sort of criminal. Writers and film-makers have produced images of the seafaring swashbuckler as being brave, ruthless and cruel with an unspecified code of honour. The image that pirates buried their treasures is not altogether true. While Captain Kidd is reputed to have hidden plunder on Manhattan Island, there are no accounts to substantiate this story or the practice. It appears that pirates divided their plunder equally when they went ashore and they gave extra shares for injuries like the loss of an eye or a limb!

Treasure Island Books

It seems that the famous writer, Robert Louis Stevenson, invented the buried treasure in his story of "Treasure Island" and subsequently caught the imagination of later writers, such as Ransome. Barrie's fantasy of pirates, fairies and eternal youth are conveyed in his classic "Peter Pan".

The Buccaneers

In the 17th century new pirates known as 'buccaneers' appeared. They were seafarers who had deserted or been marooned in the Caribbean. Many settled along the north western shores and hunted wild pigs and cattle in the savannah. From the local Indians they learned the art of 'boucan', which was the cooking of meat on a barbecue over a slow fire of dung and wood. They traded smoked meat, hides and tallow for gunpowder, muskets, cloth and spirit. The buccaneers became rich by preying on Spanish ships and ransacking coastal towns and villages such as Port Royal in Jamaica.

Sir Henry Morgan

A famous buccaneer was Sir Henry Morgan who was born in Wales and reputedly terrorised the West Indies. While besieging Portobello, he forced captured priests and nuns up ladders as human shields. On his return to London he found favour with Charles II, was knighted and died in 1688.

Captain Kidd

Pirates were normally punished by hanging at either Execution Dock at Wapping or at Tilbury. In 1701 Captain Kidd was taken to Tilbury in a cart behind a Deputy Marshall from the Admiralty Court. He was executed and then his body was tarred to preserve it as long as possible and hung in chains as a warning to any seaman contemplating piracy!

Female Pirates

Although most famous pirates were men, there were some female ones, such as Anne Bonny and Mary Read. In 1724 they were tried in Jamaica and caused a sensation when they disclosed that they had sailed in the crew of Calico Jack Rackham as men but they were in fact women and pregnant. The latter fact saved them from execution!

Ships in Museums

Many museums in London provide excellent information for students studying the National Curriculum in history, geography and technology. The collections particularly support the history study unit of Ships and Seafarers. They preserve the history of London ships and the lives of navigators who sailed in them. Several ships are also moored on the Thames. The reader may refer to the National Maritime Museum at Greenwich and the Science Museum at South Kensington for further information.

Sir Henry Morgan

Thames Barrier and Woolwich Ferry

Thames Barrier

Thames Barrier

This is the largest moveable flood barrier in the world. Completed in 1984 at a cost of £400 million, the stainless steel cowlings of the Thames Barrier stand like gateway sentinels to prevent the capital from flooding. The plan for the Barrier came about as a result of the floods of 1953 which killed 300 people downriver and threatened to engulf Central London. The disaster prompted the Government and the Greater London Council to develop plans for defending the capital from tidal surges. Half a million tonnes of concrete were poured into the blocks on the river bed as foundations for the piers and the sills on which the Barrier gates rest when not in use. When raised, each of the four main gates is as high as a five storey building and as wide as the opening of Tower Bridge. The design requirements for the scheme were to prevent surge tides passing upstream of its location, for it to be completely reliable in operation, to enable shipping to pass safely and to minimise any interference with all other aspects of the River Thames. The complete Barrier consists of four main navigation openings of 61m clear span, each containing a Rising Sector Gate to give a channel depth exceeding 9m deep.

Rising Barrier Gates

These gates are steel box girders, whose cross section is a segment of a circle. They are supported at both ends by double skin steel discs which are centrally pivoted about stub shafts projecting from the piers. The gates normally lie in the scallop of a concrete sill in the bed of the river. In operation, the gate arms rotate through 90 degrees to position the gate vertically to their flood prevention position with the high tide water level standing on the downstream of the gates. For maintenance purposes, arms may be rotated through a further 90 degrees to place the gates above water level and allow the water to move freely. The housing for the operating machinery is roofed with stainless steel. The Visitor Centre houses a collection of working models of the gates.

Woolwich Ferry

The Woolwich Ferry, a short distance downstream of the Thames Barrier, is one of the few to provide a free public service across the Thames. On 23rd March 1989, Woolwich Free Ferry celebrated 100 years of service to Londoners with a ceremony on board the Ferry's flagship John Burns. The ferry has a great history of tradition since its opening in 1889. Most of the traffic is heavy goods vehicles travelling from the south bank to the north, because of the height restrictions in the older of the two Blackwall Tunnels which takes the northbound traffic. The Captain's bridge is an octagonal cabin, about 7 feet across and is mounted above the centre of the 185 feet long ferry. The ferry is propelled by two 500 hp Lister Blackstone turbocharged diesel engines. In the centre of the ferry bridge is a square box on which two black wheels steer the direction of thrust from the vertically mounted paddles beneath either end of the vessel. The paddles are the same as those used on tugs and give the ferry the same manoeuvrability as a tug.

There are three ferries named the James Newman, John Burns and Ernest Bevin. They work eight weeks at a time making between five or six crossings every hour. The service was run by the Greater London Council until its demise in 1986 and is now under the control of Greenwich Borough Council, the operating costs are shared with Newham Council. Below deck there are wood lattice benches for passengers on foot. The ferries are among the last boats of their type afloat.

Royal Arsenal and Woolwich Dockyards

The Royal Dockyards

During the reign of Henry VIII the two Royal Dockyards of Woolwich and Deptford were established on the South East London riverside. In 1513 construction started on a new flagship, *Grâce à Dieu,* at Woolwich Dockyard for the Tudor Navy. It was a large ship, being of an estimated 1500 tons burden. A few years later a basin was built at Deptford for the maintenance and repairs of warships. Over the years the two dockyards continued to develop and Deptford became associated with distribution of stores and naval equipment. In 1607 a shipbuilding yard was leased at Deptford by the East India Company and soon the yard developed into a most extensive venture for ship building with immense stocks of timber and armaments. The East India Company was founded in 1600 and held a monopoly of overseas trade with India until 1813 and with China until 1833. It ceased to exist in 1858.

Around 1750 Woolwich was the fifth largest dockyard in the country and employed 700 men. Specialising in heavy repairs, there were a number of slipways and two docks. Deptford was the fourth largest and employed just over 800 men and included a slipway, mast houses, offices and store sheds. In 1814 the 120-gun Nelson was launched at Woolwich. It was later converted to steam. Other steam ships were built at both yards but by 1870 shipbuilding stopped and the dockyards were closed.

Deptford Creek Development

Plans are in hand for developing and transforming the Thames waterfront which stretches from Deptford in the west to Thamesmead in the east. The area covers 1000 hectares with seven miles of river frontage. It is earmarked for passenger ship facilities, a conference centre, a hotel and homes. The narrow winding alleys and creekside roads would be retained and present views preserved.

Port Greenwich Development

Further east the Tunnel Refineries, one of the largest employers locally, use the river for the import of its raw materials. This and other river industries will be retained. Plans for the development of the adjacent Greenwich Peninsular site are well advanced. British Gas owns 296 acres of the derelict land. Known as Port Greenwich, the peninsular has developed as a community of over 5,000 houses, business premises and offices, new open green spaces and community facilities.

Royal Arsenal Woolwich

Further downstream of the Royal Woolwich Dockyard was the Warren, where in 1667 Prince Rupert raised batteries on the site to provide protection to London against hostile raids from Holland. In 1671, the government purchased Tower Place and the Warren. With the sale of Artillery Gardens in London around 1682, gun proving concentrated at the Warren. In 1696 the Royal Laboratory, built for the purpose of manufacturing explosives, transferred from Greenwich. Following an explosion in May 1716 at the gun factory at Moorfields in London, it was decided to move the gun making factory out of the City and a new foundry was established next to the Royal Laboratory. Later the Royal Brass Foundry was built by Sir John Vanbrugh. The building was extensively renovated in 1974 and is now a local museum. Following a visit by George III in 1805, the Warren was renamed as the Royal Arsenal.

The Industrial Revolution, the expansion of the British Empire and the Crimean War all led to rapid expansion. At the height of World War I, there were over 72,000 employed at the establishment. It was completely self-contained with its own gas works, power station, railway, shipping service, hospital and staff quarters. During World War II, the Arsenal produced vast quantities of ammunition, but the blitz of 1940 initiated the dispersion of the Ordnance Factories and led to their eventual closure in 1967 and the disposal of much of its land to build Thamesmead. Plans for this unique historic site include a museum and heritage park. Sixteen listed buildings could become one of London's newest tourist attractions.

Rotunda Museum

Located in Woolwich Common, this museum is full of Royal Artillery guns, etc. The Pavilion design is by Nash, 1814.

The Royal Brass Foundry by Vanbrugh, restored 1974

Eltham Palace, Lesnes Abbey and Thamesmead

Eltham Palace

Eltham Palace

A 15th century Royal Palace, occupied until the reign of Henry VII with remains of more ancient Royal residences. The Great Hall with hammer beam roof is magnificent. There is a fine 14th century drawbridge over the moat. According to the Domesday book of 1086, the manor house was owned by the Bishop of Bayeux. By the end of the 13th century the Bishop of Durham presented it to Edward, the Prince of Wales, who later became King Edward II. It remained in the royal family until Elizabeth I, after which the property fell into disuse. Restoration of the hall was carried out by Stephen Courtauld in 1931-37 who lived there until 1945 when the property reverted to the Crown. The palace is about half a mile from British Rail stations of Eltham and Mottingham.

Eltham Lodge

This fine brick built house was completed for Sir John Shaw in 1664 and remained in his family until 1820. Since 1923 it has been the headquarters of the Royal Blackheath Golf Club (see page 83).

Abbey Wood and Lesnes Abbey

Located in North Bexley, Lesnes Abbey was founded in 1178 by Richard de Lucy, Chief Justice of England, for the Augustinian Canons. It was suppressed by Cardinal Wolsey in 1524 and the mansion, into which the buildings were subsequently converted, was demolished in 1844. The ruins were excavated by the Woolwich and District Antiquarian Society in 1909 and are scheduled as an Ancient Monument. They consist of the footings of walls and portions of walls in places up to 2.5m high of Kentish ragstone, flint and chalk. There are the remains of the Abbey Church and monastic buildings to the north. The "Dock", west of the Abbey remains, is probably of medieval origin. BR Abbey Wood Station is a convenient stop for a visit.

Crossness Pumping Station

Built in 1865 by J W Bazelgette, Chief Engineer of Metropolitan Works, the pumping station is at the end of the southern outfall into the Thames at Erith. It is the southern counterpart to the pumping station at Abbey Mills in Newham, north of the river. The large venetian building is in two-storeys of yellow brick; the gable ends have round arches containing round headed lights. The splendid cast iron interior houses the original beam engines by James Watt & Co. The riverside building is north of Lesnes Abbey and is served by the same railway station.

Thamesmead

Thamesmead lies on the south bank of the Thames, east of Woolwich. The Greater London Council (GLC) started to build it in the late sixties but since the abolition of the GLC it is now managed by a community based company. It was in the 16th century that the local dockyards were established at Woolwich by Henry VIII and, from the 17th century, the area was developed as a military centre. It encompassed most of Thamesmead within the Woolwich Arsenal and was used for the manufacture, testing and storage of armaments. After World War I, the Arsenal was run down because it was close to the centre of London and vulnerable to air attack. It was later bought by the London County Council and estates were completed in 1969.

Erith Museum

Located in Erith Library, the museum displays include Thames barges and a model of the "Great Harry". This spectacular warship was fitted out in 1513 at Erith's Naval Dockyard, founded by Henry VIII. BR Erith Station is a convenient stop.

Oxleas Wood

The 800 year old wood in South London is a most significant piece of ancient woodland and is a site of special scientific interest. There is a rich and diverse fauna. The woods are dominated by tall oak trees with hazel and chestnut coppices underneath. The wildlife includes hedgehogs, rabbits, foxes, squirrels, badgers and an extensive variety of resident and migrating birds. The nearest BR station is Elton Park.

Chislehurst Caves and Common

Chislehurst in the London Borough of Bromley has two main attractions; the National Trust natural woodlands and the beautifully wooded commons of Chislehurst and St Pauls Cray; and the interesting historic Chislehurst Caves opposite Chislehurst British Rail Station. Mined by the Romans and cut in the chalk hills by prehistoric inhabitants of London, there are a large number of caves linked by miles of passages, several of which are open to the public. They have had many uses including human shelter, storage and religious ceremonies. During the First World War, the caves were used as an ammunition depot and during the Second World War they formed a vast air raid shelter for Londoners. Chislehurst is easily reached from Charing Cross and London Bridge Stations.

Alexandra Palace, North London

Hampstead, Kenwood House and Keats Museum

Hampstead Village
The first mention of Hampstead, dated 986 AD, makes the village just over 1000 years old. The hilltop village grew on Hampstead Heath over a number of centuries. During the 18th century wealthy people came to the Springs of Hampstead, said to have healing powers. It has also attracted artists, painters and literary figures. The painter John Constable who is buried in the village, was a resident. His watercolours, painted at the top of the Heath, are kept at the Victoria and Albert Museum. The magnificent Heath dips into a lake and wooded valleys. Dick Turpin ranged over this Heath and the local inns were a favourite hide-out! To the north east of the park is the beautiful Kenwood House with its priceless treasures open to the public. Concert and music festivals are held during the summer evenings. Hampstead is still very much a village of attractive Georgian houses and narrow alleyways, including Church Row and Holly Mount.

Hampstead Heath Walks
Hampstead Heath is one of the finest London parks and together with its ponds and historic buildings provides an excellent green walk. Hampstead Heath Station and a car park are at the southern end of the parkland. Along the perimeter of the Heath you will find three of Hampstead's most famous public houses:

The Spaniards founded by the valet of a Spanish Ambassador, The Bull & Bush and Jack Straw's Castle. Just south of the latter is the Whitestone Pond situated near the highest point on the Heath which gives superb views of London.

Adam Bridge
A timber three span facade designed by Robert Adam, has balustrades over an ornamental pond in the Heath.

The Old Toll House
Located opposite the Spaniard's Inn off Hampstead Lane, this mid-18th century Toll House was originally at the entrance to the Bishop of London's estate which stretched eastward as far as Highgate.

Kenwood House
This fine mansion which was once the seat of Lord Mansfield, was remodelled by Robert Adam in 1764. It houses the Iveagh Bequest and includes works by many artists including Gainsborough, Turner, Reynolds, Rembrandt, Vermeer and Hals. Also displayed is a fine collection of neo-classic furniture. The house has fine plasterwork with ceiling paintings and is managed by English Heritage. Located on Hampstead Heath, Archway, Hampstead or Golders Green tube stations are nearby (081 348 1286).

Keats House and Museum
The house of the English poet, John Keats, is in Keats Grove and it was here he wrote his famous Odes. Built in 1815, and known as Wentworth Place, the property was originally a pair of semi-detached Regency villas. It was in one of these cottages that John Keats lived during the most creative period of his life from 1818 to 1820, with Fanny Brawne as his neighbour for part of that time. Part of the old basement kitchens has been restored to give some idea of life in the early 19th century (071 435 2062).

Burgh House
This fine Queen Anne house, dating from 1703, is located in the heart of old Hampstead village. It is a centre for Hampstead Museum of Local History, art exhibitions, lectures, concerts and other events. Located in New End Square, NW3, the house is near Hampstead tube station (081 431 0144).

Freud Museum
Sigmund Freud's last home is a museum of his life and great psychological contributions. It includes the famous couch and his library, antiquities and furniture brought from Vienna, Austria, in 1938. The museum is located at 20 Maresfield Gardens, Hampstead, NW3 (081 435 2002).

Kenwood House, Hampstead Heath

Highgate Cemetery and Alexandra Palace

Karl Marx tomb, Highgate Cemetery

Rose Window, Alexandra Palace

Highgate Village

Highgate Village was built up mainly during the 18th century and is one of the highest spots in London. Here you can stand level with the cross of St. Paul's Cathedral in the City. It is one of the most fascinating parts of North London.

Highgate Cemetery

In 1838 the architect and engineer, Stephen Geary of the London Cemetery Company, laid out the first Highgate Cemetery to the west of Swains Lane. Situated adjacent to Hampstead Heath, with excellent views of London, it soon became popular with the Victorians. This led to further development on the east side of Swains Lane. There are over 50,000 graves with about 166,000 bodies, including those of the famous Karl Marx, George Eliot, Dante Rossetti, Michael Faraday, John Galsworthy and Herbert Spencer. Marx was buried here in 1883 and his bust was erected in 1956. The east part is open to visitors and is a place of pilgrimage to Karl Marx. The west part can be seen through guided tours only. The architecture

of the site is a combination of Gothic and Egyptian styles which were fashionable at the beginning of the 19th century. On the north side are Egyptian Avenue and Cedar of Lebanon Catacombs which have been formed around a tree on the original estate. At the north of Lebanon Circle is the mausoleum with its fine stepped pyramid roof resembling the tomb of the Greek King, Mausolus in Halicarnassus.

Highgate Bridge

The wrought iron bridge, designed by Sir Alexander Binnie in 1897, is of segmental span resting on seven girders. Cast iron side panels have floral decoration in spandrels. There are ornamental cast iron railings and lampholders along the roadway. The piers are of Portland Stone with vermiculated rustication.

Highgate Wood

This beautiful wood, once the property of the Bishops of London, is now owned and managed by the City of London and provides a ramble leading along a disused railway

track to Alexandra Palace. It is said that Dick Whittington, three times Mayor of London, passed this wood on his way to the City. Highgate and East Finchley tube stations are nearby.

Alexandra Park and Palace

Located near Muswell Hill station, the high level park surrounding the palace has excellent views of London so take your binoculars with you. There is a small animal enclosure, a boating lake, a paddling pool and ice skating rink (081-444 7696). In the early 1930s, parts of the Palace were converted to studios by the BBC for the transmission of sound and vision programmes and all news broadcasts were transmitted from the Palace. Exhibitions were held in the main hall which was also used for various sporting functions and horse race meetings. The Palace was partially destroyed by fire in 1980 and subsequently restored to its former glory. The two massive steel box lattice arches, each spanning 52m, support a suspended 120m x 52m roof which re-creates the form of the 1875 roof destroyed by fire.

Royal Air Force Museum and Wembley Stadium

Royal Air Force Museum

Britain's National Museum of Aviation is located at Hendon near Colindale tube station. It tells the complete story of flight, the aeroplanes and the people who designed and built them. It houses the world's finest collections of over 65 aircraft all under cover. Walk under the wings of the World War II Spitfire and the famous Vulcan V bomber. Study the historical 'Battle of Britain' experience and sit in an RAF Tornado!

Harrow Public School

Harrow is one of the great public schools in the United Kingdom. Sir Winston Churchill, Lord Byron and Lord Shaftesbury are just a few of the public figures from history who were educated here. Churchill was the seventh prime minister to study at the school, and it has produced many other leaders internationally. Given a royal charter by Queen Elizabeth I, it was established by John Lyon as a free grammar school in 1572. The school buildings and chapel are on top of a 400ft hill with spectacular views.

St. Mary's Church

Nearby Harrow School is St. Mary's Church, which dates back to the 11th century. It was founded by Lanfranc, Archbishop of Canterbury in the reign of William the Conqueror and consecrated by his successor Anselm in 1094. Only the lower part of the tower and parts of the nave column remain from that period. The church was rebuilt by Sir George Gilbert Scott between 1846 and 1849. Byron's daughter Allegra is buried in an unmarked grave beneath the porch.

Church Farm House Museum

This handsome mid 17th century brick built house has a period furnished kitchen of 1820 and dining room of 1850. The museum has changing exhibitions. The building, one of the oldest in Hendon, is in Greyhouse Hill, NW4, near Hendon Central tube station and BR Hendon station (081 203 0130).

Wembley Stadium

London has a large number of sports facilities. One of the most famous sport and entertainment centres is Wembley Stadium. It has a large capacity and hosts a variety of pop and classical concerts, football matches, etc. On completion in 1924, the British Empire Exhibition was staged here. It is the annual venue of the Football Association and Rugby League Cup Finals as well as hosting international soccer matches. The 1948 London Olympic Games were also staged here. The two domes of the building are a local landmark. Wembley Park tube station is within 5 minutes walking distance. BR Wembley Complex station is in the middle of the stadium site.

Wembley Arena

Built as the Empire Swimming Pool in 1934, it was the largest covered pool in the world. The cantilevered concrete roof spans 72m (236 feet). Extensively used for concerts, conferences and exhibitions.

Wembley Stadium

Bruce Castle, Forty Hall and Epping Forest

Bruce Castle Museum

Dating from the 16th century, Bruce Castle was once the manor house of Tottenham in north London. During the 19th century the castle was used as a boarding school. Rowland Hill, the famous postal reformer, was among its headmasters. The displays include local history, postal history and the museum of the Middlesex Regiment. British Rail Bruce Grove Station is within walking distance. It is located in Lordship Lane, Haringey N17 (081 808 8772).

Forty Hall Museum

This Grade I listed building was constructed for Sir Nicholas Raynton in 1629 and it retains much of its original interior. The museum holds temporary exhibitions as well as permanent displays of furniture, ceramics and glass. It is set in beautiful rolling parkland in Forty Hill, at Enfield in north London. The British Rail Station at Enfield Chase is a convenient stop (081 363 8196).

New River

Forty Hall is within Whitewebbs Park and near the New River, financed and built by Sir Hugh Myddleton in 1609 to provide a supply of water for the City of London. The River flows from springs at Chadwell, near Ware, to the New River Head at Clerkenwell. Enfield Town by BR train from Liverpool Street is a convenient stop followed by a short bus journey to Forty Hill. There is also a car park in the grounds of Forty Hall.

Lea Valley Park & Visitor Centre

The parkland and recreation grounds stretch for about 23 miles northwards from North East London to Ware in Hertfordshire. The Visitor Centre has displays on history, countryside, waterways, events and school programmes held at the park. It is located at the junction of A121 and B194, off the Crooked Mile Roundabout (0992 713838).

Epping Forest Walks

Located within the boundaries of the London Borough of Waltham Forest the nine square miles of Royal Forest are the largest public space in the UK and are controlled by the Corporation of London. Under the Epping Forest Act of 1878, the forest was dedicated by Queen Victoria for the "enjoyment of the people for ever". The forest has a profusion of trees including the stately beech, birch, holly, hornbeam and oak. There are well known spots for ramblers such as Connaught Water, Queen Elizabeth's Hunting Lodge and High Beach. The Conservation Centre offers a range of environmental education including field studies. Lectures are held once a month on various aspects. Chingford BR station is a good place to start exploring (081 508 7714).

William Morris tapestry

Queen Elizabeth's Hunting Lodge

The museum in Ranger Road, Chingford, has a fine collection which depicts the events and daily life in the Epping Forest area from medieval times to the present day. The buildings were hunting lodges dating back to the 16th and 18th centuries. Located in the north corner of Epping Forest, the lodge was built 1543 during Henry VIII's reign. The timber-framed and L-shaped house has a spiral staircase which leads to the third floor where visitors may see the 16th century roof structure with collar beams, purlins and arched braces. The building is used as a local museum (081 508 0028).

William Morris Gallery

Along Forest Road, E17, the Water House was built in the middle of the 18th century and it contains the artist William Morris's paintings, drawings and furniture. He was born in Walthamstow in 1834 and died at Hammersmith in 1896. From 1848 to 1856 the Morris family home was the Water House. Morris's work as a designer and writer had substantial influence on the taste of the second half of the 19th century. Collections of wallpapers, printed and woven textiles, emborideries, rugs and carpets, furniture, stained glass and ceramics are on display (081 527 3782).

Hampton Court

Chiswick House and Hammersmith Mall

Chiswick House

West London

West London includes Shepherd's Bush, Hammersmith, Chiswick, Acton and Ealing. Shepherd's Bush is well known for the BBC TV studio, where audiences can watch programmes being recorded. Like the West End, Hammersmith is famous for its cinemas and dance halls such as the old Hammersmith Palais. Chiswick is an attractive riverside village where you can see the Palladian Chiswick House and the 17th century house that belonged to Hogarth. Chiswick High Street is a beautiful tree-lined wide road with fashion shops, cafes and restaurants.

Hammersmith Malls

Beside the River Thames are the Upper and Lower Malls with period houses dating back to the 18th and 19th centuries. The Riverside studios have an excellent range of facilities for art, drama and music. There are boathouses, old pubs and the Georgian house terraces include Kelmscott House.

Hammersmith Bridge

Designed by Sir Joseph Bazalgette and built 1884-87 on the piers of the old bridge of 1827. It is a suspension bridge carried by wrought iron chains with four cast iron cased towers, a cast iron arch between each pier. Monumental anchorages are cased at either end of the bridge.

Chiswick House

Chiswick House in Burlington Lane, W4, was built by the Third Earl of Burlington during 1725-30. The beautiful villa style house has a fine interior and gardens by William Kent. It was built as an extension to the old Jacobean family house which was later demolished. The design was based on Palladio's Villa Capra at Vicenza in Italy. The vermiculated Portland stone ground floor originally housed Burlington's library. The stuccoed first floor contains the state rooms arranged round a central dome, which is approached through a portico and hall. The landscaped grounds are now a public park. The Obelisk on the west side, the Deer House to the east and the Temple are part of the original design. The gateway, north west of the house, was designed by Inigo Jones and brought from Beaufort House, Chelsea in 1736. The fine bridge over the canal was built by Wyatt in 1788 (071 222 1234).

Hogarth House

William Hogarth's relics, impressions and engravings are contained in this 17th century villa in Hogarth Lane, Chiswick, W4, close to Chiswick House (081 994 6757).

Chiswick Bridge

Designed by the architect Sir Herbert Baker and the engineer A. Dryland, the bridge consists of three main segmental arches over the river with smaller arches over the two paths. It was built in 1933 of concrete faced with portland stone.

Gunnersbury Park and Museum

Once the home of the Rothschild family, the building dates from 1800 and is located in the magnificent Gunnersbury Park in West London. It is said that the niece of King Canute had a manor on the site in the 10th century. The house near Acton Town tube station, W3, displays notable items of local history and transport (081 992 1612).

Pitshanger Manor Museum

Set in an attractive park in Mattock Lane, Ealing, the manor was rebuilt 1800-3 by the architect Sir John Soane as his own home. The interior is Georgian in design. Ealing Broadway tube station is close to this local museum (081 579 2424).

Kew Palace and Royal Botanic Gardens

Kew and Richmond

Kew is best known for its Royal Botanic Gardens, containing a unique collection of trees and exotic plants. Further south is the beautiful riverside town of Richmond with its extensive Royal Park. The cricket ground on Richmond Green is overlooked by Richmond Theatre. Richmond Hill is one of the highest viewpoints in London.

Royal Botanic Gardens

Kew Gardens is one of the world's finest botanical and horticultural displays. The 300 acre gardens have many thousands of plants from all parts of the world. The primary purpose is to service the science of botany. It contains over 50,000 different trees and houses some five million dried and processed plants. It is a beautiful place in which to stroll and picnic amongst the delightful roses and colourful flower beds. Kew Gardens tube station is only a short walk away. Pleasure boats have a regular service between Westminster Pier and Kew Pier.

Kew Palace

The site of Kew Palace, or the Dutch House as it is sometimes known, was occupied in the 17th century by Samuel Fortry, a London merchant of Dutch parentage. His rebuilding of it is commemorated by his and his wife Catherine's initials and the date 1631 above the main entrance. The ornaments, pilasters and the gables with crowning pediments show extremely refined brickwork.

During the 17th century Sir Henry Capel owned Kew House and planted the first trees and herb gardens. Later the House was renamed Kew Palace and became a Royal Residence being home to Frederick, the Prince of Wales - son of George II - and his wife Princess Augusta, the mother of George III. When Frederick died in 1751, Augusta expanded the plant collection and founded a major collection of exotic species including some brought from China. She also erected Chinese follies and temples. In 1803 the old Kew Palace was demolished and the present house of the same name was built. George III and his wife lived here until her death in 1818. Today it has been restored and furnished to reflect the royal splendour of the early 19th century.

The Palm House

One of Kew's spectacular sights is the Palm House, one of the earliest buildings of iron and glass construction. Inspired by Queen Victoria, the plan for the house was presented by an Irish Engineer Richard Turner. The structure consisted originally of wrought iron ribs, curved sheets of glass, and a central dome. Building began in 1844 and took four years to complete. The interior by Decimus Burton comprises a gallery running the length of the house and contains over three thousand species of palms including an ancient Chilean wine palm. Outside displays include a rose garden, a pond and a pair of stone Chinese Guardian Lions of 18th century origin.

The Chinese Pagoda

Built in 1761 to the design of Sir William Chambers, it rises 165 ft and resides in the south west corner of the gardens. Originally it had eight glass covered dragons at the corner of each of its storeys and balconies built by Chippendale. Unfortunately now the structure is considered unsafe for visitors and is no longer open.

Queen Charlotte's Cottage

This charming two storey thatched cottage, with 47 acres of woodlands, was built circa 1770 as a summerhouse for George III's wife Queen Charlotte and was the setting of many royal tea parties. The cottage was enlarged to its present form around 1805. Queen Victoria presented the property to Kew Gardens in 1898 to celebrate her Diamond Jubilee. The cottage gives the impression that it still might be in royal use.

Kew Bridge

Built 1903, this bridge is of grey granite with rusticated rock faced voussoirs to the arches and a bold bracketed cornice surmounted by a plain parapet. The space between the arches is decorated with cartouches which contain the coat of arms of Middlesex and Surrey. The original lamp standards situated over the centre of the arches were removed in the 1960s and replaced with modern standards in the pavement. The engineer was Sir John Wolfe Barry and the architect Cuthbert Brereton.

The Palm House, Kew Gardens

Syon Manor House and Kew Steam Museum

Syon House

The splendid manor house with its magnificent interior and grounds is the home of the Duke of Northumberland and is a short distance from Kew Gardens. The approach to the house is along an avenue of old lime trees dating back to about 1700. Before you reach the house there are two stone built "Pepperpot Lodges". The whole area is surrounded by beautiful gardens and fields which slope gently to the Thames. Within the estate is the London Butterfly House.

Syon House stands on the site of a nunnery which was founded in 1415 by Henry V, and was suppressed by Henry VIII with the Decree of Dissolution signed in 1534. In 1547 Edward VI granted the estate to his uncle the Duke of Somerset, who converted the nunnery into a Tudor mansion which is the basis of the present building. Elizabeth I gave it to the Earl of Northumberland and his heirs. In 1762 Robert Adam spent several years in designing the interior of the house. During the 19th century the Third Duke rebuilt parts of the house and had the walls refaced in Bath Stone.

Grand Hall and Drawing Room

A tour of the interior starts in the splendid Grand Hall designed by Robert Adam with four Italian marble statues standing on pedestals around the room. The Dining Room has a tall clock dating from 1785 and in the Red Drawing Room, the walls are hung with crimson Spitalfields silk. The portraits in this room include those of Charles I and his wife Queen Henrietta Maria by Van Dyck, and Charles II. The print room also contains many portraits and contains some fine 18th century furniture. The house shows how members of the nobility lived two centuries ago.

The Great Conservatory

The beautiful 30 acres of grounds surrounding the house contain avenues, vistas, lawns, lakes and woodland gardens designed by Capability Brown, with many species of oak trees. The Great Conservatory is built of gun-metal and Bath stone to the design of Charles Fowler and was a model for the Crystal Palace of the Great Exhibition held in Hyde Park in 1851. There are many plants and a large display of cacti. The garden centre is one of the country's largest.

Museum and Art Centre

The Motor Museum in the grounds of Syon House with its Heritage Collection of British Cars is well worth a visit. The Art Centre has some 200 paintings on view. Visitors can reach Syon House by rail from Waterloo to Kew Bridge then a short journey by bus. Those travelling by car should go to Park

Syon Manor House

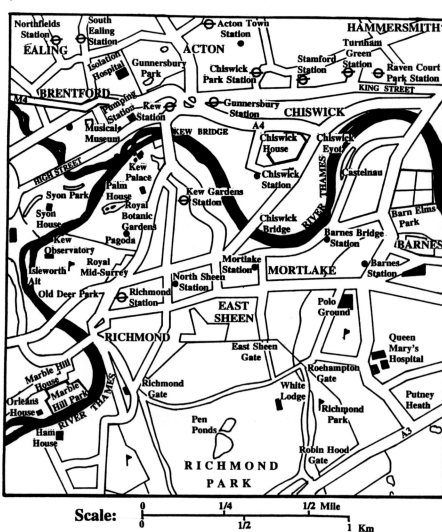

Map of Chiswick, Kew and Richmond

Road entrance. Orchestral Concerts and Grand Opera are performed in the open air during summer months.

Pumping Station Museum

This historic pumping station and monument to London's water supply system houses giant Cornish Beam engines, which were used to pump water up to service reservoirs and around London's water mains. One of these engines is steamed up on a regular basis. The museum is located in Green Dragon Lane, Brentford, just a few minutes walk from Kew Bridge BR station (081 568 4757).

Wimbledon Tennis, Common and Windmill

Wimbledon Tennis Championship

Wimbledon hosts one of the world's most prestigious sports events every year. Thousands of visitors pack into the All England Lawn Tennis and Croquet Club for the Wimbledon fortnight each summer. This is the most famous International Lawn Tennis competition in the world. It has been held on the courts of the All England Club since 1877. The Lawn Tennis Association founded in 1888, has been the governing body of the game in Britain and has shared in its organisation with the All England Club at Wimbledon. The present grounds are in Church Road on the west side of Wimbledon Park. The 16 outside grass courts are used during the championship. The seating capacity of the Centre Court is extensive. The Tournament is held from late June into early July (081 540 0362).

Lawn Tennis Museum

The museum has a fascinating collection of tennis history exhibits, many of which are gifts from great players of the past. The exhibition explains the origins of tennis and the development of the sport. Life-size models of famous players, ladies' tennis clothes of different periods and courtside equipment are shown. The library is well stocked with tennis books and periodicals. It is named after Lord Ritchie of Dundee, a former Chairman of the Stock Exchange, who took a keen interest in tennis. The museum shop has a view of the Centre Court (081 946 6131).

Wimbledon Common and Windmill

Wimbledon Common is a vast stretch of unspoiled countryside with natural woodland and scrub for rambling, horse riding and other sports. The White Windmill in the centre of the common was built in 1817 and is one of the few remaining hollow-post flour mills. It is also a local museum with collection of pictures, maps and cuttings showing the town's development from the Tudor times to the present day.

Wimbledon Common Walks

The Common has a profusion of oak, beech, hazel and silver birch trees with an undergrowth of gorse and blackberry. The Beverley Brook on the west side rises near Worcester Park and finally flows into the River Thames at Barn Elms Park. Another landmark is Caesar's Camp where a British Chieftain attempted to fight off Julius Caesar's invading army around 54 B.C. A convenient car park is at Robin Hood Gate in Richmond Park adjacent to Wimbledon Common.

Wimbledon Theatre

Wimbledon Theatre, The Broadway, Wimbledon, SW19. Opened in 1910 many famous artists and companies have performed in the theatre including the National Theatre and the Royal Ballet.

Polka Theatre for Children

In Wimbledon Broadway, the Children's Theatre has shows all year round, plus workshops, story telling, exhibitions, Young Magician of the Year and a great deal more. There are also workshops for disabled children. Copies of scripts of shows are available for loan if requested in advance.

Manresa House

Formerly Bessborough Lodge, it was built by Sir William Chambers (1762-67) at Roehampton, Wandsworth, for Lord Bessborough. Additions and alterations were made after 1860. The front of the Palladian house has five bays overlooking gardens and Richmond Park. The pedimented Ionic portico at the centre is raised on balustrades at first floor above a rusticated podium with twin curved flights with iron railings. The interior contains splendid ornamental ceilings.

Wimbledon Tennis Courts

Richmond Landmarks, Park and Riverside

Richmond Village and Riverside

Richmond has many tourist attractions and historic buildings, also there is Teddington Lock and Weir, Richmond Park, and the new riverside developments. The lock and weir form a classic piece of Victorian Civil Engineering Work and is a great place for boating enthusiasts.

The riverside development by Water Lane is a delightful area with both new and refurbished buildings providing homes and offices. The impressive buildings are grouped around the Town Hall. There is a cafe and a pub with outside seating overlooking the river. Courtyards, wide patios, walkways through landscaped lawns, planted trees and benches are well planned, making it an ideal spot for sightseers. The buildings were completed in 1989.

Historical Landmarks

The most historic part of the area is that leading up from the River by Friars Lane. The remains of the Palace of Richmond, Tudor built, is seen, including the Gatehouse bearing the Royal Arms of Henry VII. Nearby in Maids of Honour Row, dating from 1724, is a fine Queen Anne style building with cornices and inside panelled walls. Richmond Green is a beautiful green space with mature trees, a splendid place to relax and walk. It is surrounded by elegant 17th and 18th century houses, once occupied by merchants and business people who worked in the City of London. By the Green, stands the 18th century built Richmond Theatre, one of the best preserved in London.

St. Mary Magdálene Church in Paradise Road, has a Tudor Tower, with extensions dating from the 18th and 19th centuries. Also in Paradise Road, stands Hogarth House, the home of Leonard and Virginia Woolf from 1915 to 1924. It was here they started their famous Hogarth Press. St. Matthias Church on Richmond Hill is a Victorian landmark with its spire nearly 200 feet in height. From Richmond Hill you can see large houses built by the wealthy Londoners of the past. These include Ham House, Marble Hill House and Strawberry Hill House.

Richmond Park Walks

Of some 2400 acres in area, Richmond Park is second only to Epping Forest in size and is the largest urban open space in Britain. Measuring about 2.5 miles across and 2.75 miles from top to bottom, these former royal hunting grounds have beautiful trees and are ideal for walking. It contains the Pen Ponds and the wooded East Sheen Common. The White Lodge, the home of the Royal Ballet School, was the birthplace of Edward VIII and for a short period (1923-7) the residence of the Duke and Duchess of York, later King George VI and Queen Elizabeth. One of the many attractions in the park is its large herd of deer. During the summer, the Isabella Plantation has a wonderful display of azaleas, camellias and magnolias. Richmond and North Sheen are suitable stations. There is a car park at the Robin Hood Gate, off the A3.

Richmond Bridge and Pier

One of the first River Thames bridges in London, it was built 1774-77 to the design of James Paine and Kenton Couse. There are five arches in Portland stone all with balustraded parapets and it was widened in the late 1930s. The attractive bridge has been portrayed by Turner and other artists. By the bridge, skiffs and motorboats are available for hire in the summer. Richmond Landing Stage is a popular venue for pleasure boats from Westminster Pier.

Teddington Lock

The last and largest lock in the River Thames is located at Teddington and the effect of the tides reaches this far inland. From this point onwards downstream for 69 miles to the sea, the tidal Thames is under the control of the Port of London Authority. The semicircular weir was first constructed in 1912 and rebuilt after the Second World War. It has 32 radial gates which continuously measure the rate of flow in the river.

Richmond Riverside

Marble Hill, Strawberry Hill and Orleans Gallery

Marble Hill House

A Palladian villa on the north side of the Thames, Marble Hill House was built 1724-29 for George II's mistress Henrietta Howard, Countess of Suffolk, as a retreat from the court life. Here she entertained many of the writers and poets of the day including Pope and Walpole. The Great Room is lavishly decorated with gilded carvings, fine furniture, prints and paintings by Panini of views of ancient Rome. The large park surrounding the house descends gently to the river. British Rail St. Margaret's Station is within a few minutes walking distance.

Orleans House Gallery

Near Twickenham station, the remains of Orleans House now contains an art gallery and shows contemporary exhibitions with a permanent collection of paintings and prints showing local views of Richmond and Twickenham. The Twickenham villa known as Orleans House was built for James Johnson, Joint Secretary of State for Scotland under William III. The Octagon was added to the house in 1720. Built in brick and stone, it was designed by James Gibbs and richly decorated inside by the Italian plasterers Guiseppe Artari and Giovanni Bagutti. The King of France, Louis Philippe, stayed here during his exile until 1815 when he returned to France.

Strawberry Hill

Built 1749-76 to the design of Horace Walpole and a group of distinguished architects of that time, this picturesque house led a Gothic revival in England. The entrance is reached from the north past the cloister and leads into the beautiful Staircase Hall which rises to an armoury. The Library is an excellent room with a period fireplace. The spacious Gallery has fine vaulting. It is occupied by St Mary's Roman Catholic Training College and is near Twickenham's Strawberry Hill station.

Ham House

Directly facing Marble Hill House on the south bank of the river this beautiful property is surrounded by parkland. It was built in 1610 by Sir Thomas Vavasour and has remained relatively unaltered. The walls are hung with tapestries and the Great Hall contains paintings by many famous artists. Outside the magnificent 17th century garden is an aviary.

Ham and Petersham Green Walks

On the western side of Richmond Park are the delightful Ham and Petersham Commons with Ham House halfway between the two. Charles II walked his spaniels on the lawns here and the area provides excellent routes for hikers. A parking area in Richmond Park may be used as a base. Near Richmond Gate, in Petersham Meadows, there is an interesting dairy farm, the Star and Garter.

Kingston Bridge

Built 1825-28 to the design of the engineer Edward Lapidge, this fine bridge was opened by the Duchess of Clarence, later Queen Adelaide. Constructed of Portland stone, it has five rusticated arches, bold cornice and a balustraded parapet. The semi-circular cutwaters are surmounted by flat panelled piers which interrupt the balustrade. One of the original 19th century cast iron lamp standards still survives. The bridge was widened in 1914 with the old charm retained.

Riverside John Lewis Building

The Thames riverside John Lewis Store, opened in November 1992, measures 155m by 95m, the size of St. Paul's Cathedral in the City of London. One of the design considerations was the need for the Kingston Town Centre Relief Road to pass through the middle of the building. This was achieved by the construction of a bridge across the three basements, with the building superstructure forming a tunnel over the road. Other extraordinary features included the removal of the archaeological remains of a medieval vaulted cellar, which has been returned to the building for future public display. Some parts of Kingston Bridge were replaced with new construction.

Kingston Riverside with John Lewis Building

Royal Hampton Court Palace and Park

South West of London

This part of London is endowed with beautiful green open spaces including the Royal parks of Kew, Richmond, Bushy and Hampton Court plus the Common lands of Putney and Wimbledon. There are royal courts, manors and riverside houses built over a period of many centuries. The most splendid being Hampton Court on a bend of the River Thames, passed Teddington.

Hampton Court Palace

Cardinal Wolsey bought a medieval manor from the Order of St John of Jerusalem in 1514 and built the grandest mansion house which he presented to Henry VIII in 1529 in an unsuccessful attempt to retain his favour. Henry VIII added the Great Hall with its beautiful roof structure and remodelled the Clock Tower. Later Charles II built the long canal and landscaped the gardens. William and Mary transformed the Court to make it their equivalent of the French Versailles. You can visit the State Apartments with their outstanding collection of period furniture and tapestries. The beautiful gardens feature the famous Maze and the oldest tennis court in the country. The Orangeriê houses the fine series of cartoons by Mantegna depicting 'The Triumph of Caesar'.

The Tudor façade is one of the best preserved in England. The Gatehouse leads to the Base Court and then the Great Hall and Clock Court beyond. The Base Court's brickwork is richly decorated by the Italian craftsman Giovanni da Majano and is an example of Italian Renaissance fine work. Anne Boleyn's Gateway leads to the magnificent Clock Court. On the north side is the Great Hall built over cellars and completed in 1536. To the south, the grand Portland stone colonnade forms an entrance to the King's apartments.

The King's Apartments

The King's State Apartments were built in 1689 by Wren for King William and Queen Mary as part of a major redevelopment of the Palace. He demolished the older buildings of Henry VIII and created Fountain Court, the fine quadrangle around which the new apartments are grouped. Following the restoration of the fire damage of 1986, the opportunity was taken to return works of art and tapestries which had been dispersed over the centuries and visitors can now see the State Apartments much as they were when William III occupied them 300 years ago. Within the Palace are many private appartments which are leased at "peppercorn rents" to well born pensioners who have performed services to the State and Crown.

Hampton Court Park

Six hundred acres of this beautiful park surround the magnificent Royal Hampton Court Palace. The Long Water canal originated by Charles II cuts across the park and there is a minor road giving access to the area. This park and the adjacent Bushy Park are excellent for green walks. Kingston BR station is convenient for a walk to the Kingston Gate of Hampton Court Park.

Hampton Court Bridge

Built in 1933, the bridge has three arches of brick and Portland stone with a balustrade parapet. The architect was Sir Edwin Lutyens and the engineer W P Robinson.

Bushy Park

Just north of Hampton Court is another park comprising a thousand acres which is famous for its herds of wild deer and its magnificent Chestnut Avenue where 200 year old trees border the road leading to Diana Fountain. There is an Environmental Education Centre. BR Hampton Wick station is on the east side of the park and easily reached from Waterloo Station.

Hampton Court Palace

Royal Windsor Castle and Great Deer Park

St. George's Chapel and Windsor Castle

Windsor Castle

Windsor Castle, nearly a mile in circumference, is perhaps more than any other castle evocative of England and royalty. Overlooking the Thames and the eight square miles of Windsor Great Park it was built by William I as a fortress about 1070 and has been used as a royal residence since the 12th century when Henry I lived here. It was besieged in 1215 when the rebellious barons forced King John to sign the Magna Carta at nearby Runnymede. It was also used by Cromwell as a barracks. The castle was modified and extended by Sir Jeffrey Wyatville for King George IV. When the present Queen is in residence, the Royal Standard is flown from the Round Tower. At other times the State Apartments containing many treasures and paintings from the Royal Collection are open to the public. Sadly, the castle caught fire on 20 November 1992, causing extensive damage and is being renovated at an estimated cost of £100 million.

The Queen's Dolls House

Located at Windsor Castle, the Dolls House was designed by Edwin Lutyens, opened in 1926 and contains 1/12 scale models of clocks, lights, locks etc., originally belonging to Queen Mary, grandmother of the present

Queen. There are also miniature paintings and books by renowned artists and writers.

St. George's Chapel

Completed by Henry VIII, St. George's Chapel is the private chapel of the Dean and Canons of Windsor Castle and the chapel of the Order of the Garter. Beautifully carved choir stalls are overhung by the banners of the Knights of the Garter who worshipped in the chapel. The fan vaulting of the roof structure, the paintings and the iron work are splendid. Many sovereigns are buried here, including Henry VIII and his third wife Jane Seymour. The chapel was open to the public until recently when it was damaged by fire. It is said to have started with an electrical fault during maintenance work in the chapel. Plans are in hand to repair it.

Old Windsor and Wax Museum

The town of Windsor is dominated by the Castle and has a 17th Century Guildhall designed by Wren. Adjacent to the station is the Royalty and Empire Exhibition of wax tableaux depicting State events and Queen Victoria's Diamond Jubilee clebrations.

Runnymede

A meadow on the south bank of the Thames near Egham, Surrey, where on 15 June 1215

King John was forced to sign the Magna Carta, guaranteeing human rights, in reality the rights of the Barons, against the excessive use of royal power. Nearby is a memorial erected by the American Bar Association and dedicated to John F Kennedy combined with a scholarship scheme for sending students to American Universities. On Coopers Hill is a memorial to 20,455 men and women of the commonwealth air forces who died during the second World War and have no known grave.

Eton College

One of the largest and most notable of English public schools Eton was founded in 1440 by Henry VI and his statue stands in the yard. The present constitution dates from 1868 and the governing body consists of a Provost appointed by the Crown, and ten fellows. The boys number over a thousand; seventy king's scholars or collegers live in the college, the remainder called Oppidans live in masters' houses grouped around the older buildings. King George III's birthday is celebrated each year on 4 June. Distinguished old Etonians include William Pitt, the elder, (1708-78), The Duke of Wellington (1769-1852), Gladstone (1809-98), Eden (1897-1977) and MacMillan (1894-1986). The college houses the Museum of Eton Life.

St George's Hall, Windsor Castle

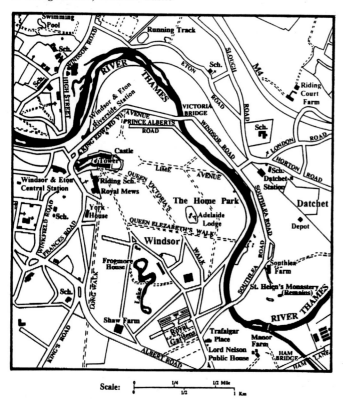

Map of Royal Hampton Court and Bushy Park

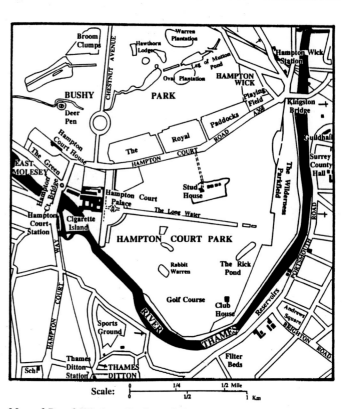

Map of Royal Windsor Castle and Great Deer Park

RIVER THAMES

(8) Steam Museum

(9) Chiswick House

(6) Syon House

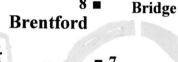

(5) Marble Hill

Kew
Bridge

8 ■

Brentford

■ 10

River

Brent

■ 7

Kew Palace
& Gardens

6 ■

■ 9

Chiswick
Mall

Chiswick
Bridge

Chiswick
Eyot

Mortlake

Beverley Brook

Twickenham
Bridge

Richmond Bridge

Richmond Hill

Crane

5 ■

4 ■

■ 3

2 ■

Eel Pie
Island

(2) Strawberry Hill

Teddington
Lock

(3) Ham House

River
Thames

Kingston
Bridge

(1) Hampton Court Palace

Kingston Museum

Hampton Ct.
Bridge

■ 1

Hampton Court
Park

Hogsmill

Surbiton

Thames Ditton

Great Riverside Attractions

(1) Hampton Court Palace - East Molesey, Surrey (081 781 9500).

(2) Strawberry Hill - St Mary's College.

(3) Ham House - Ham Street, Richmond, Surrey (081 940 1950).

(4) Orleans House - The Riverside, Twickenham, Middlesex (081 892 0221).

(5) Marble Hill - Richmond Road, Twickenham, Middlesex (081 892 5115).

Richmond Riverside & speciality shops - South of River Thames.

(6) Syon House - Syon Park, Brentford, Middlesex (081 560 0881).

(7) Kew Palace & Gardens - Kew, Richmond (081 940 1171).

(8) Kew Bridge Steam Museum - Green Dragon Lane, Brentford (081 568 4757).

(9) Chiswick House - Burlington Lane W4 (081 995 0508).

(10) Hogarth House - Hogarth Lane, W4 (081 994 6757).

Gunnersbury Park Museum - Acton W3 (081 993 1612).

(11) Fulham Palace - Bishops Avenue, Fulham, SW6 (071 736 5821).

(12) Chelsea Harbour - London SW10

(13) Chelsea Royal Hospital - Chelsea, SW3 (071 730 0161).

(14) Battersea Park & Power Station - SW19 (081 871 7530).

(15) Tate Gallery - The Millbank, SW1 (071 887 8000).

Gunnersbury Park Museum

Queens Gallery

Charing Cross

Waterloo
Bridge

Hungerford
Bridge
(Railway)
■ 19

Westminster
Bridge 17 ■
16 ■

Lambeth
Bridge

15 ■

(10) Hogarth House

Army Museum

(17) Houses Of Parliament

Vauxhall
Bridge

Albert 13
Bridge ■

Chelsea
Creek
■
12

Chelsea
Bridge

River
Thames

Hammersmith
Bridge

(12) Chelsea Harbour

■ 14

Battersea
Bridge

Beverley
Brook

Fulham Palace
■ 11

Putney
Bridge

Wandle

Wandsworth
Bridge

(14) Battersea Power Station

(19) South Bank Art Centre

(16) Historic Westminster Abbey - SW1 (071 222 5152).

(17) Houses of Parliament - SW1.
Victoria Embankment - WC2.
Courtauld Institute Galleries - Somerset House, WC2 (071 873 2526).

(18) Lambeth Palace & Gardens - SE1
Museum of Garden History - Lambeth Palace Road, SE1 (071 261 1891).

Florence Nightingale Museum - 2 Lambeth Palace Road, SE1 (071 620 0374).

(19) South Bank Arts Centre - SE1. Near Waterloo tube station (071 921 0600).

Museum of The Moving Image - South Bank, SE1 (071 928 3535).

(20) St Paul's Cathedral - Ludgate Hill,

The City, EC4 (071 248 2705).

(21) Southwark Cathedral - London Bridge, SE1 (071 407 2939).

Kathleen and May Schooner - Cathedral Street, SE1 (071 403 3965).

Gabriel's Wharf Market - 56 Upper Ground, Waterloo, SE1.

Shakespeare Globe Theatre - Bankside, Southwark, SE1.

(22) Monument - EC3 (071 626 2717).

(23) Tower of London - Tower Hill, EC3 (071 709 0765).

Tower Hill Pageant - Tower Hill, EC3 (071 924 2465).

Tower Bridge - SE1 (071 403 3761).

(24) Hays Galleria Tooley Street, SE1 (071 403 4758).

(25) HMS Belfast Ship Museum - Anchored off Hays Galleria (071 407 6434).

(26) Design Museum - Butlers Wharf, SE1 (071 403 6933).

(27) St Katharine Docks - Wapping, E1

(28) Tobacco Dock & Pirate Ships - Wapping, E1 (071 702 9681).

(29) Greenland Dock and Marina - Surrey Docks, SE16 (071 252 2244).

(30) Canary Wharf - Isle of Dogs, E14.

(31) London Docklands Development Corporation Visitor Centre - Limeharbour, E14 (071 512 3000).

(32) National Maritime Museum - Romney Road, SE10 (081 858 4422).

Cutty Sark - King William Walk, SE10 (081 858 3445).

Royal Observatory - Greenwich Park, SE10 (081 858 4422).

(33) Thames Barrier - Unity Way, SE18 (081 854 1373).

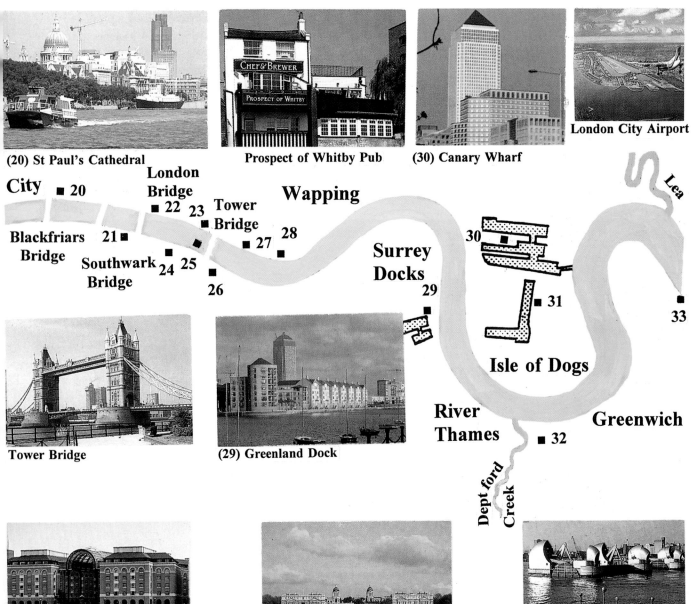

(20) St Paul's Cathedral

Prospect of Whitby Pub

(30) Canary Wharf

London City Airport

City ■ 20

London Bridge ■ 22 23

Tower Bridge ■

Wapping

Blackfriars Bridge 21■

Southwark Bridge 24 25

26

■ 27 28

Surrey Docks 29

30

■ 31

33

Isle of Dogs

River Thames

Greenwich

■ 32

Deptford Creek

Lea

Tower Bridge

(29) Greenland Dock

(24) Hays Galleria

(32) Royal Greenwich

(33) Thames Barrier

Riverside Pubs & Restaurants

Bishop Out of Residence Thames Street, Kingston, Surrey (081 546 4965) BR Kingston.

White Cross Riverside, Richmond, Surrey (081 940 6844) BR Richmond.

Dove Inn 19 Upper Mall, Hammersmith, W6 (081 748 5405).

Ships Inn Ships Lane, Mortlake, SW14 (081 876 1439).

Dukes Head 8 Lower Richmond Road, Putney, SW15 (081 788 2552) Tube East Putney.

The Ship 42 Jews Row, Wandsworth, SW18 (081 870 9667) BR Wandsworth.

Ovations Restaurant South Bank, SE1 (071 928 2033) Tube Waterloo.

Review Restaurant Level 3, Royal Festival Hall, South Bank, SE1 (071 921 0800) Tube Waterloo.

T S Queen Mary Victoria Embankment, WC2 (071 240 9404) Tube Embankment.

Savoy River Restaurant Strand, WC2 (071 836 4343) Tube Embankment.

Founders Arms Bankside, SE1 (071 928 1899) Tube London Bridge.

The Anchor Bankside 34 Park Street, SE1 (071 407 3663) Tube London Bridge.

London Docklands For waterside pubs in Docklands please see page 69.

The River for Environmental Studies

The River Thames is a rich subject for environmental studies. Its port history, commerce, industry and wild life provide excellent material for project work and study. Special boat charters and organised cruises for parties of up to 100 passengers may be arranged for schools, colleges and societies.

Thames Birds and Wildlife

The best area for birds is Rainham and Dagenham marshes east of London, where many varieties of breeding and visiting birds can be seen. These include some thousands of teal from Scandinavia with hundreds of widgeon and pintail from Russia which move between the lagoons and the river. There are numerous other birds including reed and sedge warblers and yellow wagtails. Across the river at Dartford Marshes similar species may also be seen. West of London between Chiswick and Teddington there are several wooded islands, known as aits or eyots, which provide winter grounds for cormorants and roosting sites for herons. Isleworth Ait is a dense woodland managed by the London Wildlife Trust. The Oliver's Island is just downstream of Kew Bridge, and is a densely wooded habitation with many birds.

Discover The River Thames by Special Cruises

Sightseeing by river boat

River Thames Boat Services

The River Thames is the artery of London and is a familiar feature of some of its famous places. You may know it as the setting for the Oxford and Cambridge Boat Race, for regattas or the old Doggett's Coat and Badge Race. Maybe you use it for fishing, canoeing or rowing; to walk along its banks or simply enjoy its waterside pubs and watch the world go by! But there is a lot more to discover about the Thames. There are many crafts operating up and down the river between Hampton Court and the Flood Barrier. Enjoy the circular cruises, Sunday lunch and evening cruises, some entertaining passengers with floodlit suppers and discos.

The summer services outlined below commence early April and normally continue until the end of October. During the winter season there are usually no upriver services and fewer evening trips on the pleasure boats. Please contact the relevant boat operator for departure times and types of services.

Downriver Services

Westminster to Tower - Boat Trips Limited (071 515 1415).
Tower to Westminster - Tower Pier Launches (071 488 0344) 30 minute trip.
Westminster to Greenwich -Thames Passenger Boat Services (071 930 4097). Trip 40-50 minutes every 30 minutes. Service from Greenwich (081 858 3996).
Westminster to Barrier - London Launches (071 930 3373) and Tidal Cruises (071 839 2164). 75 minutes each way.

Charing Cross to Tower - Catamaran Cruisers (071 839 3572) 20 minute trip.
Tower to Greenwich - Catamaran Cruisers (081 858 3996). 50 minute trip.
Tower to HMS Belfast - Livetts Launches (081 468 7201).
Greenwich to Barrier - Campion Launches (081 305 0300).

Upriver Services

Westminster to Kew - Westminster Passenger Services Association (WSPA) (071 930 4721) 1½ hour trip.
Westminster to Richmond - WSPA (071 930 4721) 2½ to 3 hours.
Westminster to Hampton Court - WSPA (071 930 4721) 3 to 4½ hours.

Evening and Special Cruises

Luncheon Cruise from Westminster - Catamaran Cruisers (071 839 3572) 2 hours.
Floodlit Supper Cruise from Westminster - Catamaran Cruisers (071 839 3572) 1½ hours.
Disco Cruise from Westminster - Catamaran Cruisers (071 839 3572) 4 hours.
Evening Cruise from Westminster - Westminster Passenger Services Association (071 930 2062) 45 minutes.

Special Functions

The following operators offer boats for hire for private lunches, dinners, dances, discos and conferences.
Admiral Enterprises 331 Main Road, Westerham Hill, Kent TN16 (071 237 3108).

Woods River Services P.O. Box 177, Blackheath, SE3 9JA (071 481 2711).
Voyager Club 6th Floor, 99 Kensington High Street, W8 5ED (071 937 8923).
Tidal Cruises Westminster Pier, Victoria Embankment, SW1A 2JH (071 928 9009).
George Wheeler Launches Limited Westminster Pier, Victoria Embankment, SW1A 2JH (071 930 1616).

River & Canal Boat Operators

Capital Cruises Riverboats for sightseeing and charter. 31 Mendip Road, Battersea, SW11 3SF (071 350 1910).
The Elizabethan Twin decked Mississippi paddle steamer, 99 Kensington High Street, W8 5ED (071 937 7994).
Jason's Trip Canal cruises from Little Venice to Camden Lock and London Zoo. 60 Bloomfield Road, Little Venice W9. (071 286 3428).
London Waterbus Company Canal boats from Little Venice to Regents Park and London Zoo. Camden Lock, NW1 8AF (071 482 2323).
Parkway Launches One hour circular cruise from Westminster Pier. 14 Burdell House, Parkers Row, Dockhead, SE1 2DH (071 237 1322).
Riverbus Thames Line Ltd Regular passenger service from Chelsea Harbour to London City Airport (071 512 0555).
Thompson Launches Ltd Pleasure boats from Westminster Pier to the Tower and Greenwich or upstream to Kew. 24 Walton Road, Surrey KT8 (081 979 2642).

Boat Races and Trip from Westminster to Tower

River Thames Annual Events

Barge Driving Race
The annual Greenwich to Westminster Barge Driving Race produces one of the fastest races on the river when the crews compete along the 7½ mile course.

Doggett's Coat and Badge Race
The race was first staged in 1715; Watermen and apprentices under the age of 26 are allowed to compete in the event. The race covers the four miles and five furlongs from the site of the Old Swan at London Bridge to the Swan Inn at Chelsea. The winner is presented with the Doggett's Coat and Badge at a lavish City dinner. It is normally held during the last week of July.

Oxford and Cambridge Boat Race
Excellent views of the race can be seen from a number of public houses along the course including the Dove Inn at Hammersmith and the Ship Inn at Mortlake. Around 3pm in the afternoon of the last Saturday in March, the world famous race sets off from the University Stone by Putney Bridge and the two crews row flat out for four and a half miles to Mortlake. Thousands of people line the river banks and millions more watch the race on television. What started in 1829 as a private contest between friends at the two universities has become a great annual event.

The Great River Race
This is one of London's most colourful river events, covering a 22 mile marathon course to find the traditional boat champions. The race starts from Ham House, Richmond, and finishes at Island Gardens, Isle of Dogs.

Boat Trip to The Tower
The starting point for your journey is Westminster Pier with the majestic Houses of Parliament and Big Ben in the background. Adjacent is Westminster Bridge built in 1862 to replace an older stone one of 1746. On the south bank of the River is the old County Hall, which until 1986 was the headquarters of the GLC, and is now being redeveloped as a hotel complex.

Charing Cross and the South Bank
Further downstream you pass under Charing Cross Railway Bridge with the new complex over the railway tracks on the north side of the river. The footbridge alongside is a popular route to the South Bank Arts Centre. Further along the Victoria Embankment is Cleopatra's Needle, erected in Egypt circa 1500 BC and brought to London in 1819.

Waterloo and Somerset House
The pleasure boat continues under the fine Waterloo Bridge which opened in 1945 replacing an older one of the mid 19th century. Famous landmarks on the north of the river are Somerset House and the Doggett's Coat and Badge public house.

Blackfriars and St. Paul's
Passing under Blackfriars Bridge, opened by Queen Victoria in 1860, you will see on the North Bank the Mermaid Theatre and St. Paul's Cathedral, which was the scene of the wedding of the Prince and Princess of Wales.

Cannon St. and Southwark Cathedral
The next bridge is Southwark Bridge, opened in 1921 replacing an older built in 1819. Further along is Cannon Street Railway Bridge which was opened in 1866 and renovated during the 1980s. On the south side of the river is Southwark Cathedral dating back to the 13th century.

London Bridge and Hays Galleria
Next is London Bridge, a modern structure completed in 1973 to replace one by Thomas Telford of 1831. An even earlier bridge which opened in 1209 was the only bridge across the river until 1750. Passing under the bridge on the north side you can see the Monument, built in 1677 to commemorate the Great Fire of London. On the south side is the impressive London Bridge City with its fashionable shopping precinct Hays Galleria.

The Tower and Tower Bridge
Opposite is the Tower of London, the most ancient of London landmarks dating back to William the Conqueror. Here are kept the Crown Jewels and the Tower ravens, all of which are guarded by the Beefeaters (Yeomen of the Guard) in their ceremonial costumes. Facing you at Tower Pier is Tower Bridge, the most famous bridge in the world. The bridge and its two Gothic towers were completed in 1894 and house the machinery for raising the bridge. The upper walkway across the length of the bridge is open to visitors, as is the Bridge Museum on the south side of the bridge.

Attractions Around Tower Pier
The River Bus Tours depart and arrive from this pier which is also a regular berth for the Royal Yacht when visiting London. When you leave Tower Pier, there are a wealth of attractions nearby to visit such as the London Pageant, Tobacco Dock Shopping Precinct, HMS Belfast Ship Museum and the London Dungeon, near London Bridge Station. I hope you enjoy your trip along London's broadest highway, Old Father Thames!

Sailing Barges Race from Greenwich to Westminster

London's Grand Union and Regents Canals

Grand Union & Regents Canals

The Grand Union Canal is one of the most important and interesting pieces of industrial heritage in Britain. In 1793 the 'Canal Age' was at its height and 24 Acts for Canal Construction were passed by Parliament. The Grand Junction Canal Company built the 93 miles from Brentford to Braunston around 1800. William Jessop was the civil engineer in charge. The Grand Union was created in 1932 by the amalgamation of eleven independent companies.

The Regents Canal along the northern edge of Regents Park connects the Grand Union Canal to Limehouse Basin and the River Thames in East London. During 1993, the Grand Union Canal celebrated its 200th anniversary. The towpath alongside the canal has been cleared so that walkers can make the 140 miles trek from Little Venice in Central London to Birmingham's Gas Street Basin. It is estimated to take about 14 days. Since the heavy commercial canal traffic between London and the Midlands ended in 1970, it is now heavily used during the summer by leisure craft. The canal connects to the River Thames at Brentford in West London and at Limehouse in East London Docklands. The narrow canal boats are brightly painted in traditional designs.

Camden Lock

A pleasant water area of small craft workshops and studios with two restaurants and a weekend market all situated along Regent's Canal in Camden Lock Place, NW1. Explore London's famous canal, stroll down the tow paths and catch a glimpse of the wildlife. It makes an ideal and fascinating walk. There are boat trips from Camden Lock, off Chalk Farm Road, to London Zoo or Little Venice or whole day trips to London Docklands (071 482 2250).

Canal Walks in London

It is possible to walk some 40 miles along the towing path from East to West London as shown by the map on this page. From Ladbroke Grove westwards, the public has free access to the path. East of this section the relevant London Borough has responsibility for locking access during the hours of darkness but the path is normally open from 9am till dusk. The canals are also rich in perch, pike, roach and tench; a permit is required for fishing. The following canal walks enable you to see interesting sights and to discover items of industrial archaeology. There are many pubs along the way and the nearest stations are shown. For safety, please be accompanied along these walks.

(1) Commercial Road Lock to Camden Town (6 miles in 3 hours).
(2) Camden Town to Little Venice (2½ miles in 1¼ hours).
(3) Little Venice to Willesden Junction (3¼ miles in 1¾ hours).
(4) Willesden Junction to Bull's Bridge (9½ miles in 4/5 hours).
(5) Brentford Lock to Hayes, Middlesex (5 miles in 2½ hours).
(6) Hayes to Uxbridge (5¼ miles in 2½ hours).

London Canal Museum

12-13 New Wharf Road, Kings Cross, London N1 (071 713 0836). The museum tells the story of London's canals over the past 200 years from their early days as important trade routes to today's leisurely pursuits. You can learn about the cargoes, types of craft, the barge people and the horses that pulled the boats.

Further Information

For further information and free leaflets, please contact:
British Waterways, The Toll House, Delamere Terrace, Little Venice, London W2 6ND (071 286 6101).
Port of London Authority Devon House, London E1 (071 265 2656). The PLA provides information on the River Thames.

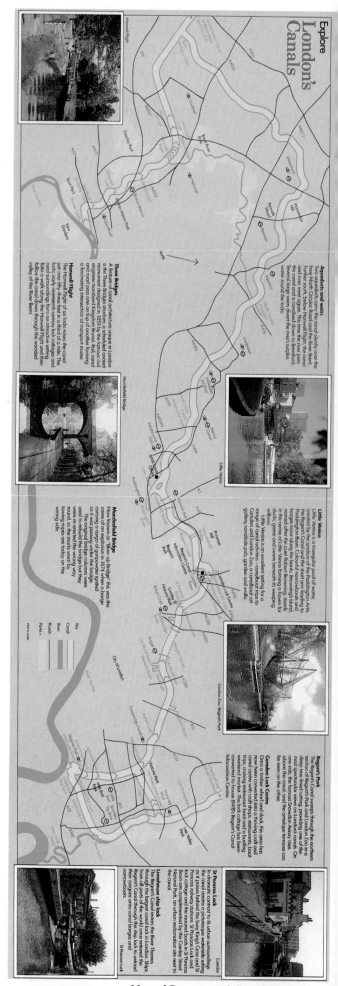

Explore London's Canals

Map of Regents and Grand Union Canals

TRANSPORT AND LONDON INFRASTRUCTURE

Expansion of London

Although London was already beginning to spread by 1830 the biggest expansion took place in the second half of the 19th century. By 1900 successive building booms had transformed London into a metropolis which continued to absorb the surrounding villages. This expansion was the result of new means of transport such as trams, railways, omnibuses and cabs. The boom in public transport during the early 20th century, hastened the astonishing growth of the London Boroughs and the suburbs and created one of the biggest metropolitan transport networks in the world.

London Railways

The earliest passenger railway in London ran south east from London Bridge on a four mile viaduct to Greenwich. The project was privately financed and cost £400,000. Construction started in April 1834 and the first section from Deptford to Greenwich was opened in February 1836. The line was extended to London Bridge by 1837. While fares were relatively high, one shilling for first class and sixpence for third class carriages, the trains ran every quarter of an hour. Within ten years, two million passengers were using the railway each year. The London and Birmingham railway terminated in 1837 at Euston Station, the Great Western in 1838 at Paddington and the London and Croydon branched from New Cross Gate in 1839. Railway construction continued during the second half of the 19th century. The North London was built to carry cargo and coal from the West India and Poplar Docks to Camden Town in 1852. In 1864 the Great Eastern reached Liverpool Street Station. Few further steam railways were built after this, the last being the Great Central line to Marylebone in 1898. London's railways today are extensive, covering virtually every part of the capital. The express InterCity network connects all major UK cities with London.

Transport in the 20th Century

The beginning of the 20th century marked the start of the motor car and by 1914 practically all buses and lorries were petrol driven and the horse drawn carriages were disappearing from the streets of London. It also signalled the building of highways and later the motorways. Today's M25 ring road was in fact conceived by the London County Council in 1943. The formation by British Rail of its Network South East was to cater for the expansion of the commuting area of London to places such as Peterborough, Norwich, Rugby, Swindon and Gloucester. Although the motorway era began in the mid 1950s and rapidly led to the construction of over 5000 miles of highways radially from London, there has been very little penetration into the centre of the city. The improvement in the means of transportation in the Capital is central to the development of trade, and generally to the enhancement of the quality of life of Londoners.

Docklands Light Railway

Opened in 1987, the Docklands Light Railway (DLR) the newest railway in Britain, is a good way to explore Docklands. It is a scenic ride taking you high over the docks, among the new buildings and through the interesting parts of the East End. The Island Gardens terminal is close to the entrance of the foot tunnel to Greenwich and the Maritime Museum. This line is being extended to Lewisham.

London Buses

An excellent way to travel and see sights is from the top deck of the buses which pass by most of the capital's landmarks, great shops and railway stations. Special All Night Buses run through Central London, serving Piccadilly Circus, Leicester Square, Victoria and Trafalgar Square, convenient for cinemas and theatres.

Air Transport

The rise in air travel did not occur until after 1945 with the construction of Heathrow and Gatwick. Mass transport became possible with the development of the Comet aircraft in 1952 and the Jumbo Jet in 1968. London City Airport in the Royal Docks was created in 1987. London's third airport in Essex, Stansted, was recently completed in 1992.

Heathrow Airport

London's status as an international world city in the post war years was strengthened by the opening of Heathrow Airport in West London in 1946. Development has proceeded over the past three decades including the addition of Terminal 4 in 1986. Earlier in 1977 the Piccadilly Underground Line was extended to the heart of the airport thus improving links with central London. Today, it is one of the busiest airports in the world with Terminal 4 alone handling up to 4,000 passengers an hour travelling to and from overseas.

Kings Cross and St Pancras Railway Stations

London's Subterranean World

Subterranean London

London has an extensive subterranean infrastructure including crypts, tunnels, pipes, sewers and cables, some with little or no written documentation. Some of these features date back many centuries but most of them were built from the middle of the 19th century onward. The subterranean world of services such as gas, electricity, water, sewerage, telecommunications, and transport systems remain a mystery to most Londoners and visitors. In this section the history and development of mains services and infrastructures are briefly outlined.

Underground Tube Lines

London had the world's first underground railway The Metropolitan Line, which ran four miles from Paddington to Farringdon Street. Opened in 1883, it took three years to construct and was an instant success; within six months 26,000 people were using it daily. In order to relieve traffic congestion within the City of London, the underground railway was expanded to join the major stations. The Circle Line followed in 1884. The District and Metropolitan Lines expanded soon after from the Circle Line into the suburbs and helped considerably in new housing developments and the establishment of Greater London. This period also marked the introduction of electric trains to replace steam. Tunnelling techniques were also improved which resulted in more tubes. The City and South London was followed by the Waterloo and City in 1898, the Bank to Shepherd's Bush 1900, Baker Street and Waterloo 1906, Piccadilly and Brompton 1906, and Charing Cross, Euston and Hampstead, 1907. No other tubes were built until the Victoria Line to Walthamstow 1971 and the Jubilee Line 1979.

The underground, or tube, is the easiest and fastest method of transport around the city. In Central London you are always within a few minutes walk of a station. After 9.30 a.m. you can buy a cheap Travel Card for unlimited travel by tube, train and bus for a day. (See inside front cover).

Post Office Underground Trains

The Post Office operates electric underground railways which run between sorting offices and British Rail main line stations. The trains are fully automated and driverless.

London's Main Sewers

Following the Public Health Act of 1858, Sir Joseph Bazalgette, the Chief Engineer of the Metropolitan Board of Works devised a system for sewage disposal in London which is still in operation. Five main sewers, three north of the Thames and two to the south, carried the sewage over 10 miles to Barking and Plumstead. One of the conduits is part of the Victoria Embankment which

Bazalgette also designed. Here 27 acres of land was reclaimed from the Thames reducing its width and removing the mudflats which were mainly deposits of sewage discharged into the river at that time and resulted in bad smells in the area. The massive retaining wall stretched for one and a third miles from Westminster to Blackfriars in a gentle curve along the line of the Thames. The Victoria Embankment transformed the north bank of the river, relieving congestion in the Strand, and becoming a pleasant place for a stroll or rest with fine views and public gardens.

London's Water Supply

Early water supplies to London were either from wells or rivers. The capital had been built on water bearing gravel beds and springs abounded. The rivers Walbrook and Fleet (now underground) brought supplies to the City from the hills to the north. During the 13th century, conduits were constructed to bring water from outlying springs to points around the City. General distribution was by means of water carriers known as cobs. In 1581 the first water wheels at London Bridge were used to pump water. Early in the 17th century water was carried to London along a canal, called the New River, which followed the Lea Valley and still operates today. By 1850, eight companies supplied Greater London through an extensive network of underground pipes. They merged in 1903 to form the Metropolitan Water Board.

In 1974 a major reorganisation took place when England and Wales were divided into ten major regions each under a Water Authority. Thames Water Authority took over the management of the whole of the Thames Valley from the Cotswolds to Gravesend and became responsible for water supply, drainage and sewage disposal. During 1992, the authority was privatised and now functions under various companies.

New London Water Ring Main

The new London ring main is being constructed as a 50 mile tunnel from Ashford and Sunbury in West London to Coppermills at Walthamstow in the North East. The huge tunnels run 120 feet below ground and when completed in 1996 will supply half of London with water. Substantial extensions are proposed during the early 21st century.

Brunel's Thames Tunnel

The year 1993 was the 150th anniversary of the opening of Sir Marc Brunel and his son Isamband Brunel's Thames Tunnel. When opened in 1843 the pioneering tunnel had taken 18 years to complete. It was the first river tunnel in the world and was sold in 1865 to the East London Railway Company. Today it still carries the trains of the underground East London Line. There is an excellent exhibition at the Brunel Engine Museum in Rotherhithe.

Blackwall Tunnel

The 1.4 km northbound bore of the Blackwall Tunnel was opened in 1897 and the southbound in 1967. The traffic flow for the first year's operations was almost 10,000 vehicles a day, mostly horse drawn wagons. Today nearly 100 years later the northbound traffic is over 40 times that figure.

Dartford Tunnel

First opened in 1963, the tunnel forms the northbound river crossing of the M25 motorway on the east side of London.

Underground view of tube lines at Bank Station

Subterranean Visits and Trips in London

Brunel's Thames Tunnel, completed 1843

Brunel Thames Tunnel Walk
The Brunel Museum may arrange dinner for a group in the Engine House, Rotherhithe, followed by a guided walk through the Thames Tunnel from Rotherhithe to Wapping and back, after the last train of the East London line runs on Saturday night. The walk will finish about 3am on Sunday morning. Contact (081 318 2489).

London Dungeon
Beneath London Bridge Station, the museum illustrates tortures, executions and murder in life size tableaux (071 403 0606).

Trip on Mail Railway
Running at 20 to 30 metres underground from Mount Pleasant sorting office and control room to Paddington. Contact the Post Office Railway (071 637 1022).

Tour of New British Library
Visit of basements excavated to nearly 23 metres to house 12 million books. Contact the Press Office (071 323 7791).

Tracking Fleet River
A walk along the course of North London's most famous underground river, which runs from Hampstead and Highgate to Blackfriars. Contact Platform (071 408 3738).

Disused Station Walk
Walk through Down Street Station, closed in 1932 and converted to the Railway Executive Committee Headquarters during World War Two. Contact (071 637 1022).

Tracking the Walbrook River
A walk along the course of Walbrook River running underground in the City of London. Contact Platform (071 403 3738).

Tracking the Effra River
A walk along the course of South London River Effra that falls from Norwood, through Brixton and discharges into the Thames at Vauxhall. Contact Platform (071 403 3738).

Visit to the London Water Main
Descend down the London Water Ring Main shaft, to the 40 metre deep tunnel at Battersea. When completed by the end of the century it will supply more than half London's water needs. Contact Thames Water (071 837 3300) or (071 637 1022).

Visit to Electricity Substation
London Electricity primary underground substation under Leicester Square in the West End was opened by the Queen on 4th June 1992. It operates inaudibly and unmanned with access through a hydraulic pavement hatch (071 637 1022).

Visit to Cabinet War Rooms
Sir Winston Churchill's wartime underground room under Whitehall in King Charles Street, SW1, with entrance from Horseguards Road. The museum is open daily (071 930 6961).

St. Bride's Church Crypt
World War II bombing of St. Bride's Church in Fleet Street revealed the remains of successive places of worship dating back to Roman times (071 353 1301).

Security Archives Eisenhower Centre
A walk through a tunnel complex near Goodge Street, W1, and 45 metres below street level. It was built in 1941 as an Air-raid shelter beneath the Northern Line and used by General Eisenhower as his operational base from 1944. Currently, it provides secure storage for films, videotapes and computer discs (071 637 1022).

Visits to Archaeological Sites
Visits may be arranged to any currently accessible archaeological excavations in Central London dating back to Roman times. Contact Museum of London (071 972 9111).

Visit to Chislehurst Caves
Mined by the Romans and Londoners, there are a large number of chalk caves linked by miles of passages, several of which are open to the public. BR Chislehurst station is opposite the caves (081 467 3264).

Dark Ride Museum
London's first dark ride museum with automated cars that travel in time, taking visitors past scenes depicting the development of London from Roman times to the present day. Contact Tower Hill Pageant (071 709 0081).

Hays Galleria Cellars
Enjoy fine wine and seafood in the bar and restaurant of the cellars of this Victorian wharf conversion at London Bridge City, Tooley street, SE1 2HN (071 403 4758).

Guildhall Crypt
This 15th century crypt with a fine vaulting is the most extensive medieval crypt in London. Visitors wishing to see it should contact the Beadle (071 606 3030).

London Historic Railway Stations and Hotels

Charing Cross Station and Hotel

Charing Cross Station was built as the terminus of the South Eastern Railways on the site of Hungerford Market. The Hotel above the Station was designed by E M Barry and built 1863-64 and was one of the first major buildings in London to be faced with artificial stone.

Paddington Station and Hotel

Paddington Station is one of the oldest surviving Victorian train sheds and is a Grade I listed building. Designed by Isambard Brunel and completed in 1854, the station served as a terminus for the Great Western Railway with connections to Bristol and South Wales. The roof structure consists of a series of wrought iron arch ribs at 3m centres. Longitudinal iron lattice valley girders are supported at about 9m centres on cast iron columns 6m above platform level. The ornamental parts of the three old sheds are by M D Wyatt. The hotel on the south side of the station is a separate building by Phillip Hardwick.

Euston Station

In 1837 Euston became London's first main line terminus with trains going to and from Birmingham, the second largest city in the UK. There was an impressive approach with a Doric style arch while in the great hall, decorated with statuary, a magnificent staircase led to the Shareholders Room. The architect for the station was Phillip Hardwick.

St. Pancras Station

The Victorian station sheds and buildings were built 1868-74 with extravagant Gothic ornaments for the hotel. The shed, designed by the engineer W H Barlow is one of the largest to be built, with a span of 243ft (73m) and a length of 690ft (210m).

Midland Hotel at St Pancras

The Grade I listed 400 bed Midland Hotel at the front of St. Pancras Station, is entirely different from the station's shed. It was designed by Sir George Gilbert and completed in 1873. It is a mainly red brick high Victorian building, incorporating a rich variety of Lombardian and Venetian medieval characteristics with features resembling English and French cathedrals. Groups of towers, turrets, gables, bays, balconies, statues and pinnacles combine to produce a building much admired nationally. The Midland stopped functioning as a hotel in 1935. The exterior is being renovated for the possible conversion of the building into a five star hotel by the end of the 20th century, which would be ideally located for the new Channel Tunnel terminal at St Pancras.

King's Cross Station

The Victorian Station was built 1851-2 to the design of Lewis Cubitt. It has two large stock-brick arches that close the ends of the sheds towards the Euston Road. It serves railway lines to northern cities and Scotland.

Liverpool Street Station and Hotel

One of the major railway stations in London, it was built in 1875 to serve as the metropolitan terminus for the old Great Eastern Railway and served trains from north east London. Today it serves East London, Essex, East Anglia and more recently Stansted Airport. The hotel was originally designed by Charles Barry and completed in 1884. With the Bishopsgate developments taking place either side of the station, the whole area has become a showpiece of modern transportation infrastructure.

Victoria Station

The Victorian Station is the amalgamation of what was originally two separate parts. To the east side are two fine vaulted iron and glass sheds built for the London, Chatham and Dover Railway Station platforms 1 to 8. On the west side is the London, Brighton and South Coast Railway Station (platforms 9 to 17) which was rebuilt 1898-1908.

Grosvenor Hotel

The impressive hotel is a reminder of the British Empire and the Victorian age. On the first and top storeys, between the arched windows, are medallions with portraits of Queen Victoria and Prince Albert.

Orient Express

Victoria Station is the main South of England rail terminal and it is the London terminal of the Orient Express which made its first trip in 1833, travelling from Paris to Constantinople. In 1977 it made what was believed to be its last run before being revived and renamed Venice Simplon Orient Express by James Sherwood, President of Sea Containers. In May 1982, over 500 bottles of champagne were consumed by the 160 passengers and this record still stands today.

Waterloo Station and Terminal

Waterloo is an important Southern Region Station, handling many hundreds of trains daily. It was first opened in 1848 and rebuilt in 1922 when it was opened by Queen Mary. The undulating and tapering glass and steel roof of Britain's first modern international rail passenger terminal is one of the newest landmarks in London. The huge terminal, completed in 1993, has an arched glass roof.

Liverpool Street station

21st Century Jubilee Extension and Cross Rail

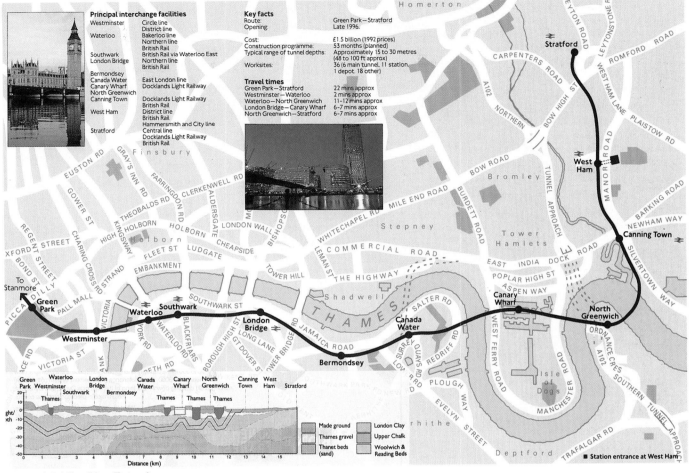

Map of Jubilee Line Extension

Jubilee Line Extension

The Jubilee Line Extension is one of the largest and most prestigious construction projects of the 1990 decade. It will cost £2bn - £3bn. Improved transport between the West End, Docklands, South East and North East London are the main areas to benefit but, by opening up new business and employment prospects, it will have a much wider impact on the future of London and transport in the 21st century. It involves expanded stations, new stations and a depot, 12.4km of twin tunnel at 30 metres deep from Green Park to Canning Town, and 3.6km of surface railway along existing British Rail North London line. The 10-mile extension has been designed to take 36 trains per hour when complete in 1997/98.

Of the 12 stations on the line, Southwark, Bermondsey, Canada Water, Canary Wharf and North Greenwich will be completely new. Station platforms in the new tunnel section will have screens between train and platform with doors synchronised with the train doors as a safety feature. The tunnels will be larger than existing tube tunnels with a side walkway to enable emergency services to gain quick access. All new stations will have fire-hardened lifts which will mean wheelchair users will be able to use the tube.

It will take 22 minutes to travel from Green park to Stratford and at peak period it is expected up to 18,000 passengers will travel in each direction. The extension has been designed to carry 30,000 passengers each way to cater for future growth. There will be 36 construction sites comprising six main tunnel sites, 11 station sites, one depot at the old Stratford Market and 18 others. Much of the spoil from tunnelling will be removed by barge to cut lorry movements. A Code of Construction Practice has been drawn up for contractors to ensure minimum disruption.

21st Century Cross Rail

The Cross Rail project involves the construction of twin rail tunnels deep under Central London. This should relieve congestion on the most crowded sections of the Central Line, Metropolitan Line, Eastern Suburban Services and the main line stations at Paddington and Liverpool Street. The service will run from Aylesbury and Reading in the west to Shenfield in the east and will incorporate part of the Underground services between London, Amersham and Chesham. Completion is planned for the year 2000.

Map of proposed Cross Rail

London Orbital Motorway M25

M25 Orbital Motorway

The M25 London Orbital Motorway is the longest city bypass in the world. Fully opened in 1986 it provides direct motorway access to all the UK's major industrial, production and distribution centres, linking with 32 radial motorways and trunk roads. The motorway, circling the Capital at a radius between 13 and 22 miles (21-35 Km) from Charing Cross, is 117 miles (188 Km) long. The total cost of construction was over one billion pounds. For the road surface more than 2 million tonnes of concrete and 3.5 million tonnes of asphalt were used. Among its civil engineering structures are the two tunnels in the north eastern quadrant: the Holmesdale Tunnel at Waltham Cross (junctions 25 and 26), and the Bell Common Tunnel at Epping (junctions 26 and 27). The Epping Foresters Cricket Club, founded in 1947, plays its matches on top of the latter tunnel. More than 2 million trees and shrubs have been planted along the whole length of the motorway. The capacity of the M25 is being increased rapidly from three to four lanes each way by 1995 and possibly to seven by the beginning of the 21st century.

Queen Elizabeth II Bridge

Completed in October 1991, this bridge is the southbound route of the M25 over the Thames at Dartford and is the first privately financed infrastructure scheme in the United Kingdom. It consists of a four-lane high level road bridge of 'cable stay' construction, with a main span 450 metres in length, the longest such structure in Europe. The consortium who built the bridge also purchased the existing road tunnels which form a vital link in the M25 crossing of the Thames on the east side of London.

Tourist and Visitor Destinations

The M25 is a valuable tourist asset. The map below shows a selection of destinations and airports which may be more easily reached by the M25 than driving through London. They include a range of attractions such as abbeys, country parks, historic houses, sporting venues and museums. Service stations are available at Junctions 2, 12 and 21.

Map of Orbital M25

(1) Forty Hall (2) Lea Valley Park (3) Waltham Abbey (4) Epping Forest (5) North Weald Airport (6) Thames Barrier Visitor Centre (7) Brands Hatch Motor Racing Circuit (8) Knole (9) Chartwell (10) Biggin Hill Airfield (11) Horniman Museum (12) Crystal Palace Sports Centre (13) Epsom Racecouse (14) Box Hill (15) Polesden Lacey (16) Chessington - World of Adventure (17) Wimbledon (18) Clandon Park (19) Wisley Gardens (20) Sandown Park Racecourse (21) Hampton Court Racecourse (22) Kempton Park Racecourse (23) Twickenham   (24) Thorpe Park - Adventure Park (25) Ascot Racecourse (26) Kew Gardens and Palace (27) Heathrow Airport Visitors Gallery (28) Colne Valley Park (29) Windsor (30) Cliveden (31) Bekonscot Model Village (32) Wembley Stadium Complex (33) RAF Museum (34) Aldenham Country Park (35) St Albans (36) Hatfield House

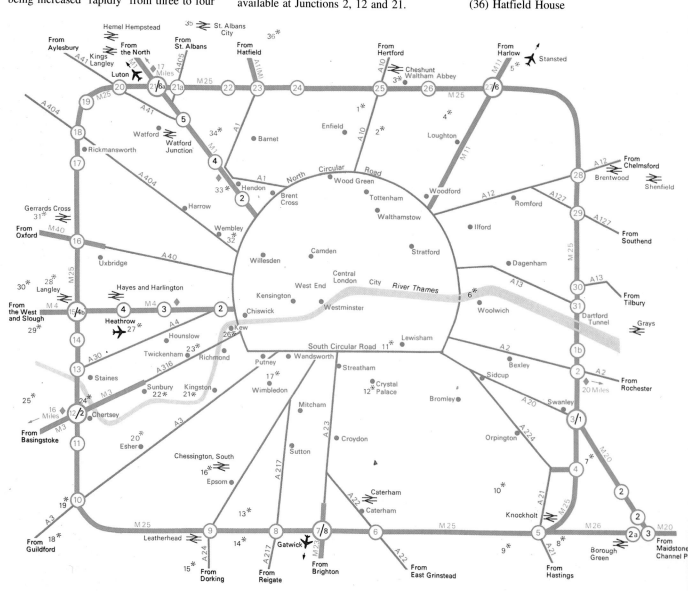

London to Europe - The Channel Tunnel

Channel Tunnel interior

Travelling to Europe

For Londoners and tourists the first obstacle in travelling to Europe is crossing the Channel. The Channel Tunnel which opened in 1994 provides a vital link in the European transport system by connecting, for the first time, Britain's road and rail network to those of France and continental Europe. Eurotunnel is an Anglo-French private sector company which in 1986 signed the Channel Tunnel Concession with the British and French governments for the design, construction and operation of the tunnel system for 55 years.

Eurotunnel shuttle trains transport cars, coaches, motorcycles and lorries and their passengers swiftly between the two terminals in Folkestone and Calais, affording direct access to the French and English motorway systems. High speed passenger trains operate a three-capitals service, between London St. Pancras Station, Paris Gare du Nord Station and Brussels Midi Station. During the peak periods, the frequency of shuttles is every 15 minutes in the day and hourly at night. The journey times from London will normally be just over 2 hours to Brussels and Paris, 6 hours to Milan, 8 hours to Berlin and 11 hours to Copenhagen. No booking is necessary for the shuttle trains, they operate 24 hours a day.

Channel Tunnel Construction

The Channel Tunnel is one of the largest European civil engineering projects this century. The two operating tunnels, each 7.6 metres diameter, contain a single railway track. At intervals of about 375 metres they are linked by cross-passages to a central service tunnel. These passages reduce the build up of air pressure in front of the moving trains. The linings of all three tunnels are either constructed from cast iron or reinforced concrete segments. Each of the main tunnel segments weighs 10 tonnes and cement grout was pumped behind the linings to provide a final seal with the rock. Comprehensive drainage, ventilation, signalling, control and monitoring equipment is installed to operate the service at maximum 300 shuttles daily. The 50km long tunnels are divided into three approximately equal sections by cross over points. This makes it possible to isolate parts of the tunnels for night maintenance or for any emergency, while operating a reduced single-line service.

The tunnel system runs between 25 and 45 metres beneath the sea bed, the level being determined by a stratum of chalk marl through which it was bored. By staying almost entirely within soft rock, normally impermeable to water, the boring machines worked rapidly and safely. Above the chalk marl is a layer of grey chalk, a harder but more fractured formation. Beneath it is gault clay, which is impermeable to water.

Channel Tunnel Rail Link

Union Railways, part of British Rail, is developing the Channel Tunnel Rail Link from Dover to St Pancras Station. The preferred route runs from St Pancras alongside the North London line to a tunnel from Dalston to Barking. The tunnel from King's Cross to Barking is also an option. Passing over the Dartford Tunnel and under the Queen Elizabeth II Bridge, the line dips steeply into a 3.2km tunnel under the Thames. At Gravesend a junction can be built to discharge international traffic cargo and let domestic traffic proceed to London. A bridge carries the line across the River Medway. From there a 4km tunnel under the North Downs in Kent takes the line from the M2 to the M20 corridor. A 200m cut and cover tunnel is constructed at Boxley. The route bypasses Ashford and then follows parallel to the M20 to Dover. The total length of the line is 108km. An international station would be built at Ashford. A station at Stratford would consist of a reinforced concrete box 1km long. Other possible stations north of the Thames are at Meadway Parkway or at Ebbsford to serve Dartford and Gravesend with a junction for the North Kent line. The detailed route for the £2bn - £3bn project has been finalised and completion is expected early in the 21st century.

Paris Euro Disneyland

Euro Disneyland is a magical world of fun, fantasy and adventure. Its attractions are set in five theme 'lands'. They include Main Street, USA, which represents 19th century small town America and hosts a parade of Disney Characters. Frontierland has the big Thunder Mountain roller coaster train ride. Adventureland offers fun on the high seas while Fantasyland is a true fairyland. Discoveryland features breathtaking special effects. All attractions except the shooting gallery are included in the entrance ticket price and there are special rates for children aged 3-11 years. The holiday resort also contains six hotels and a camping ground.

QUICK GUIDE TO LONDON

Royal Kensington Palace and Gardens

Royal London: Palaces, Castles and Yachts

Royal Yacht Britannia

The Albert Memorial Kensington Gardens, SW7. The impressive 53 metre high memorial was commissioned in 1872 by Queen Victoria in memory of her husband, Prince Albert, who died in 1861. The frieze around the base depicts 169 life size figures of artists and men of letters.

Banqueting House Whitehall, SW1. Built 1619-21 for James I, it has magnificent ceiling paintings by Rubens and was a major place for Court entertainment.

Buckingham Palace Westminster, SW1. It has been the home of English Kings and Queens since 1837.

Changing of the Guard Buckingham Palace, SW1. The Queen's Guard is changed in front of the Palace at 11.30.am daily.

Royal Greenwich and Observatory Greenwich, SE18. The birth place of Henry VIII, it was first converted into a Royal Hospital for seamen and then a Royal Naval College. Through the Observatory passes the zero meridian.

The Crown Jewels Tower of London, E.1. The Crown Jewels, including the huge cut diamond 'The Star of Africa' are on view to visitors, - Captain Blood attempted to steal them in 1671!

Kensington Palace Kensington Gardens, SW7. Queen Victoria was born here in 1819 and stayed until she became Queen in 1837 when she moved to Buckingham Palace. Today, Princess Diana, wife of Prince Charles, and her children live here. There is a 'Royal Robe' Museum.

Kew Palace Kew, Middlesex. George III and Queen Charlotte used the palace during the first half of their reign.

Hampton Court Palace Twickenham, Middlesex. It remained a royal palace from 1526 in the days of Henry VIII until the time of George II, and is now partly occupied by 'Grace and Favour' pensioners.

Queen Charlotte's Cottage Kew, Middlesex. This charming thatched cottage was built in 1771 as a summer house for Queen Charlotte.

Royal Britain Exhibition Aldersgate Street, EC2. Situated opposite the Barbican tube station the museum tells the story of 51 Kings and Queens who ruled Britain from 973AD to the present day.

Royal Mews Buckingham Palace Road, SW1. It contains the Queen's horses and carriages including the Gold State Coach used for coronations since 1820.

Royal Yacht Britannia The Royal Yacht commissioned in 1959 contained some furniture from the previous Victorian yachts and is used by the Queen to visit Commonwealth and foreign countries.

Royalty & Empire Museum The visit of Queen Victoria to Windsor and Eton Central Railway Station in 1897 has been re-created by Madame Tussaud in a special exhibition in the Royal Waiting Room.

The Royal Nore This is the royal barge on the River Thames, owned by the Port of London Authority. When used by Her Majesty the Queen she is attended by her Bargemaster and Seven Royal Watermen.

State Opening of Parliament The Queen rides from Buckingham Palace to the Palace of Westminster for the opening of Parliament.

St George's Chapel Windsor Castle, Bucks. It has been the private chapel of the Dean and Canons of Windsor and also the Chapel of the Order of the Garter.

St James's Palace Whitehall, SW1. This was built in 1532 and became a royal palace in 1698 when Whitehall was destroyed by fire. The Chapel Royal has been the setting for many royal marriages.

Tower of London Tower Hill, E1. Built by William the Conqueror at the end of the 11th century, the White Tower has been a fortress, royal residence and prison.

Trooping the Colour Horse Guards Parade, Whitehall, SW1. A spectacular military pageant is held annually on the second Saturday in June to honour the sovereign's official birthday. As Colonel in Chief, the Queen takes the salute at the Horse Guards Palace where the Brigade of Guards and the Household Cavalry are paraded.

Westminster Abbey City of Westminster, SW1. The Abbey has been used for the coronations of monarchs since the crowning of William I in December 1066.

Westminster Hall City of Westminster, SW1. Built by William Rufus, son of William the Conqueror, the 72 metre long hall has a magnificent hammer beam roof.

Windsor Castle and Deer Park Windsor, Berks. Built first as a fortress by William the Conqueror, it is the most famous royal residence after Buckingham Palace.

The Yeoman Warders Tower of London, E1. Created by Henry VII in 1485, the sovereign's personal bodyguard wears a colourful state dress uniform which has remained unaltered since that time. The famous Ceremony of the Keys has taken place at the Tower every night for centuries. The keys are then locked away in the Governor's office for safe keeping.

London Great Art Galleries and Top Places

Great Art Galleries

Barbican Art Gallery, Level 8 , Barbican Centre, EC2Y 8DS (071 638 4141). The Gallery is a major venue for changing exhibitions. Tube Barbican or Moorgate.

Courtauld Institute Galleries, Somerset House, Strand, WC2R ORN (071 873 2526). The Galleries house the internationally famous Courtauld Collection of impressionist paintings. Tube Temple or Covent Garden.

Design Museum, Butlers Wharf, Shad Thames, SE1 2YD (071 403 6933). The temporary exhibitions include the Mae West Lip Sofa of the 1920s and the prototype of the 1992 Olympic Gold Medal winning bicycle. Tube London Bridge or Tower Hill.

Hayward Gallery, Belvedere Road, The South Bank, SE1 8XZ (071 261 0127). The gallery mounts temporary exhibitions of contemporary and historical art. Tube Waterloo or Embankment.

Horniman Museum, London Road, Forest Gate Hill, SE23 (081 699 2339). The museum has exhibitions relating to cultures of the world. BR Forest Hill.

National Gallery Trafalgar Square, WC2N 5DN (071 839 3321). The gallery houses the permanent National Collection of European Paintings. Tube Trafalgar Suquare.

Maritime Museum, Romney Road, Greenwich, SE10 (081 858 4422). The museum contains the world's greatest collection of art and artifacts, including boats and naval uniforms. BR Greenwich, or by river trip to Greenwich Pier.

National Portrait Gallery, St. Martin's Place, WC2H OHE (071 306 0055). The Gallery has a collection of paintings of famous British men and women. Tube Leicester Square.

The Queen's Gallery, Buckingham Palace, SW1A 1AA (071 799 2331). The Gallery houses the Royal Collection. Tube St. James Park or Green Park.

Royal Academy of Arts, Burlington House, Piccadilly, W1 (071 439 7438). The permanent collections include paintings, sculpture, prints and drawings. Tube Piccadilly Circus or Green Park.

Tate Gallery Millbank, SW1 (071 887 8000). British and foreign paintings of all periods. Admission free. Tube Pimlico.

Victoria & Albert Museum, Cromwell Road, South Kensington, SW7 2RL (071 938 8500). The Museum contains the world's finest decorative art treasures, including furniture. Tube South Kensington.

Wallace Collection, Hertford House, Manchester Square, W1 (071 935 0687). The house has a permanent exhibition of paintings, French furniture and armour. Tube Bond Street or Baker Street.

Whitechapel Art Gallery, 80 Whitechapel High Street, E1 7QX (071 377 0107). The Gallery exhibits modern and contemporary art especially by artists working in East London. Tube Aldgate East.

Top Places to Visit

The following top tourist attractions are arranged in alphabetical order.

British Museum Great Russell Street, WC1, Tube Russell Square. Archaeological and historical relics. Admission free.

Buckingham Palace The Mall, SW1 Tube Green Park. Official residence of Her Majesty the Queen. Admission Charge.

Canary Wharf Isle of Dogs, E14. DLR Canary Wharf. One of the world's biggest office complexes with the tower at 800 ft dominating the skyline. Admission free.

Covent Garden WC2 Tube Covent Garden. This fashionable area has specialist shops, old pubs and opera house.

Houses of Parliament Westminster, SW1 Tube Westminster. The House of Commons and House of Lords. Admission free on appointed days.

Madame Tussaud's Marylebone Road, NW1. Tube Baker Street. World-famous waxworks museum. Admission charge.

National Gallery Trafalgar Square, WC2 (see above).

St. Paul's Cathedral Ludgate Hill, EC4 Tube St. Paul's. The City's great cathedral. Admission fee.

Tate Gallery Millbank SW1 (see above)

Thames Barrier The eighth wonder of the modern world built to save London from flooding. Boat trip to Greenwich Pier. Admission Charge.

Tower of London Tower Hill, EC3. Tube Tower Hill. World-famous medieval castle and buildings. Admission charge.

Victoria and Albert Museum South Kensington, SW7. Tube South Kensington Treasure house of art. Admission free.

Westminster Abbey Parliament Square, SW1. Tube Westminster. Kings and Queens have been crowned and buried here. Admission charge.

The Tate Gallery

London Sights for Free and Highest Viewpoints

London Hilton and Hyde Park Corner

Sights with Free Admission

There are many attractions in Central London with free admission; here are some excellent examples that must not be missed.

British Museum, This is still the biggest attraction in London. Its treasures range from the Elgin Marbles to a 5000 year old Egyptian mummy (071 636 1555).

Bank of England Museum There are showcases displaying phases in the Bank's history.

Guildhall This is the medieval seat of power in the City of London. The Lord Mayor of London, in full regalia, holds Court of Common Council in the Great Hall each month. The ceremony is accompanied by Sword Bearer, Mace Bearer and City Marshall (071 606 3030).

St. Paul's Cathedral Sir Christopher Wren's masterpiece has the second largest church dome in the world (071 248 2705).

Stock Exchange The heart of the old trading floor in the City where colourful computer screens flash the ups and downs of the Financial Times (FT) Index. Book in advance for a lecture (071 588 2355).

Lloyds of London Visitors are stunned by the new skyscraper next to Leadenhall Market (071 623 7100).

National Gallery (see page 118).

Wallace Collection (see page 118).

Sotheby's See exhibits on view before they go under the hammer at the world's oldest art auctioneers (071 493 8080).

Burlington Arcade A Dickensian atmosphere awaits you at one of London's oldest shopping malls in Mayfair.

Tate Gallery (see page 118)

South Bank Art Centre This has the world's greatest galaxy of galleries, concert hall and theatre.

Old Covent Garden A bustling area of Central London with its shops, arcades and glassblowers at the Glasshouse in Long Acre.

Sir John Soane Museum This personal collection remains as it was when the architect Sir John Soane was alive during the 18th century (071 405 2107).

Hyde Park See the many sculptures including the statue of Peter Pan. Behind this statue hidden in the shrubbery, are small gravestones that symbolise the many fairies from John Barrie's novel, but you'll have to look very hard!

London's Highest Viewpoints

Permission is required before entering these buildings.

Alexandra Palace On Muswell Hill, about 250 ft, North London, N22. Nice views.

Canary Wharf Tower There are fine views from this 800ft high building. Viewing is not allowed at the present time.

Centrepoint At New Oxford Street, 372 ft high. Lift and two galleries are available.

Hampstead Heath Constable's famous view of London can be seen from the 450 ft high point on the Heath.

Jack Straws Castle NW3 Views across London from a terrace in North End Way, NW3.

London Hilton Roof bar at a height of 325 ft in Park Lane. Fine views of Hyde Park, Buckingham Palace and Mayfair.

British Telecom Tower At 500 ft it gives one of the highest views in London but is unfortunately closed at the present time. It also has the highest revolving restaurant.

Shell Building In Belvedere Road, SE1, the 317ft building gives fine views over the River Thames and London.

Nat West Tower At 183m high this is the second highest building in London.

St. Paul's Cathedral Has magnificent views of London, its churches, the Tower, and the River. There are 727 steps leading to the highest point of 335 ft.

The Terrace, Richmond Hill In Richmond, Surrey, pleasant views from a height of 180 ft over west and south London.

Westminster Cathedral Ashley Place, SW1, 270ft high, it has fine views of London and Westminster.

Tower Bridge Walkway
The bridge walkway about 140 feet above high water gives superb views of the City.

English Heritage Historic Houses and Palaces

Greater London

Chiswick House, Burlington Lane, W4 (081 995 0808). One of the first Palladian villas built circa1725 in London by the architect Lord Burlington. The beautiful interior decoration is by William Kent. There is an exhibition and film telling the story of the house and gardens. BR Chiswick and Tube Turnham Green.

Eltham Palace North of A20 off Court Yard, SE9 (071 222 1234). The delightful feature of this 13th century palace is the Great Hall with its splendid hammer-beam roof. BR Eltham or Mottingham.

Jewel Tower, Westminster. The Victoria Tower, opposite South end of Houses of Parliament, SW1 (071 222 1234). Built circa 1365 to house the personal treasures of Edward III, it was formerly part of the Palace of Westminster. It housed valuables which formed part of the King's wardrobe and subsequently used as documents and government office. There is a new exhibition entitled "Parliament Past and Present." Tube Westminster and BR Charing Cross.

Kenwood, The Iveagh Bequest, Hampstead Lane, NW3 (081 348 1286). Standing on Hampstead Heath, Kenwood House contains the most important private collection of paintings donated to the nation. There is a self portrait by Rembrandt and paintings by British artists such as Turner and Reynolds. BR Hampstead Heath.

London Roman Wall, Near Tower Hill underground station, EC3 (071 222 1234). The best preserved part of the old Roman Wall, raised in the Middle Ages, that formed part of the eastern defences of the City of London circa AD200.

Marble Hill House Richmond Road, Twickenham (081 892 5115). A magnificent Palladian villa built 1724-29 and set in 66 acres of parkland along the north side of the River Thames. The Great Room has beautiful gilded decoration and paintings by Panini. BR St Margarets and Tube Richmond.

Ranger's House Chesterfield Walk, Blackheath, SE10 (071 222 1234). A fine red brick villa built circa 1700 on the edge of Greenwich Park. It contains a number of Jacobean portraits and a collection of musical instruments. BR Maze Hill.

Winchester Palace, Southwark Near Southwark Cathedral, at corner of Clink Street and Storey Street, SE1 (071 222 1234). This is the remains of the Great Hall of this 13th century town house of the Bishops of Winchester, badly damaged by fire in 1814. BR and Tube London Bridge.

Westminster Abbey Chapter House. Approach either through the Abbey or through the Dean's Yard and the cloister. Adjacent to Houses of Parliament, SW1 (071 222 5152). The Chapter House, built by the

Kenwood House Library

royal masons circa 1250 and restored during the 19th century, contains the finest examples of medieval English sculpture. The 11th century Pyx Chambers today houses the Abbey treasures. The Abbey Museum contains medieval royal effigies. Tube Westminster and Victoria, BR Charing Cross.

Around London

Audley End House West of Saffron Waldon on B1383 from M11 exit 8 (0245 492211). The Jacobean Mansion is set in magnificent parklands landscaped by Capability Brown, through which flows the River Cam. There are thirty rooms on view, including reception rooms by Robert Adam, with fascinating collections of paintings and period furniture. BR Audley End.

Farnham Castle Keep North of Farnham on A287, M3 exit with A3 (0483 605757). Built as a fortified manor by the medieval Bishops of Winchester, this motted and bailey Castle has been occupied since the 12th century. The historic town of Guildford is nearby. BR Farnham.

Tilbury Fort East of Tilbury off A126, Essex (0268 541662). The largest and best preserved fort of 17th century military engineering in England. It commands the River Thames and shows the development of fortifications during the 18th and 19th centuries. The exhibitions demonstrate how London was protected from seaborne attacks. BR Tilbury.

London Colourful Ceremonies and Events

London Events

London is famous worldwide for traditional and ceremonial activities such as the Changing of the Guard, Trooping the Colour, State Opening of Parliament and the Lord Mayor's Show which attract large numbers of spectators and tourists. These daily, annual and unique events are held at historical locations. With so many exhibition halls and sports arenas, there is also a packed programme of nationally popular events throughout the year. These include the Proms in the Albert Hall, Henley Regatta on the Thames, the Royal Tournament at Earls Court, the Horse of the Year Show and the Battle of Britain celebrations are just a sample of these popular annual events.

Daily Colourful Ceremonies

The Changing of the Queen's Guard
At Buckingham Palace, the New Guard, following the band, arrives from Wellington Barracks for a ceremony lasting about half an hour. Not held in bad weather.

The Changing of the Queen's Life Guard Held at the Horse Guard Arch, Whitehall. The ceremony known as Changing of the Guard lasts about 20 minutes starting at 11 am. The Guard is also inspected on foot at 4.00 pm.

Ceremony of the Keys Held at the Tower of London, the Chief Warder with an escort from the Brigade of Guards, locks the west middle tower and Byward tower gates at night with traditional ceremony.

Annual Events

January
International Boat Show Earls Court, Warwick Road, SW5 (071-385 1200). Early January.

Royal Epiphany Gifts Chapel Royal, St. James's Palace, Marlborough Road, SW1. Held at 11.30am on 6 January, this picturesque ceremony is centred on two Gentlemen Ushers offering gold, frankincense and myrrh on behalf of the Queen.

England v Wales Rugby Match Whitton Road, Twickenham, Middlesex (081-546 2081). Late January.

Racing Cars Show Olympia, Hammersmith Road, W14 (071-603 3344).

February
Cruft's Dog Show Olympia, Hammersmith.

English Folk Dance and Song Festival Royal Albert Hall, South Kensington, SW7. (071-589 8212).

March
St. David's Day Royal Windsor Castle, Berkshire. Leeks are given to the Welsh Guards at a ceremony on the 1 March, normally attended by the Duke of Edinburgh.

Ideal Home Exhibition Olympia, Hammersmith. Sponsored by the Daily Mail newspaper, from 5th to 30th March.

St. Patrick's Day Pirbright. A colourful ceremony on 17 March in which shamrocks are given to the Irish Guards.

Spring Antique Fair Chelsea Old Town Hall, Kings Road, SW3. (071-352 8101).

Oxford v Cambridge Boat Race The River Thames race from Putney to Mortlake, the last Saturday in March.

April
London Marathon 3rd Sunday in April (081 948 7935). International race.

Chelsea Flower Show, Chelsea Royal Hospital Late April (071 834 4333).

May
Football Association Cup Final Wembley Stadium (071 262 4542).

Royal Windsor Horse Show 2nd week in May (0298 72272).

June
Beating Retreat, Horse Guards Early June, **colourful event**.

Derby Day United Racecourses (0372 726311). National horse race.

Trooping the Colour, Horse Guards Second Saturday in June

Royal Ascot 20 June (0990 22211).

Wimbledon Tennis Tournament End of June/beginning of July.

July
Henley Regatta at Henley, Bucks. Rowing extravaganza along the Thames, early July (0491 572153).

Barge Driving Race 22 July Greenwich to Westminster (081 930 0907).

Doggett's Coat at Badge Race 25 July, London Bridge to Chelsea (071 626 2531).

Royal Tournament at Earls Court End of July (071 287 0907).

The Proms at the Royal Albert Hall July-September.

August
Notting Hill Carnival, Carribean Mardi Gras in London, August Bank Holiday

September
The Derby, Coronation Cup at Epsom Race Course Early September.

The City of London Flower Show 6th and 7th September at Guildhall.

Battle of Britain Celebrations Middle of September

October
Horse of the Year Show at Wembley Arena First week of October (0298 72272).

November
Lord Mayor's Show and Procession Early November (071 606 3030).

London / Brighton Veteran Car Run First week of November

London Film Festival Middle to late November, South Bank, SE1 (071 928 3535).

December
Royal Smithfield Show at Earls Court. First week in December (071 235 7000).

Showjumping Championship Olympia (071 373 8141).

Model Engineering Exhibition at Olympia - Late December.

The Lord Mayor's Show

London Walks, Festivals, Radio and TV Shows

The BBC, Langham Place

Walking Tours

For those looking for active and low cost view of the City, then walking tours are certainly very attractive. Pub tours usually end in a social gathering.

Capital Walks 44 Kent Avenue, Ealing, W13 (071 924 2071). Guided walks of the City of London, Samuel Pepys' London, Legal London and Clerkenwell village.

Citisights of London 145 Goldsmiths Row (071 739 4853). Tours to learn about the history and archaeology of London.

City Walks 9/11 Kensington High Street W8 (071 937 4281). Guided London walks.

Cockney Walks 32 Anworth Close, Woodford Green Essex (081 504 9159). Daily guided walks through East London.

Discovering London 11 Pennyfields, Warley Brentwood Essex (0277 213704). Walking tours of pubs, museums, etc.

Footloose in London P.O Box 708, NW3 (071 435 0159). Private guided walks from Hampstead Tube Station.

Guided Walks of Historic London 3 Hamilton Road Golders Green NW11 (081 455 7542). Walking tours with sights of historical interest.

London Walks P O Box 1708, London NW6 (071 624 3978). Extensive range of walks including Hampstead Village.

Pub Walks The walks last about 2 to 3 hours. There are normally two stops at pubs along the way for refreshment with food available at some pubs (071 485 6415).

Jack The Ripper Murder Trail

The trail with a guide starts from Whitechapel tube station and tours the murder territory. Just over 100 years ago in the autumn of 1888, a reign of terror gripped Victorian London when a series of five horrific murders were committed in the area. Suspects included a mad doctor and an heir to the throne. Within three months the murders stopped. The Ten Bells public house is in the middle of Ripper territory, where you can see old newspaper cuttings on the pub walls or sample the Ripper Tripple, a red cocktail kept in a dusty green bottle behind the bar (081 668 4019).

Annual Art Festivals

Various festivals are held in London during the summer months. Details may be obtained from the organisation listed or from the London Tourist Board on (071 730 3488).

Camden Festival Music, Opera, choral and symphony concerts are performed with exhibitions, poetry and drama. Four weeks from late February. Information from Camden Town Hall (071 837 7070).

City of London Festival Music festival with choirs, orchestras, and concerts at St. Paul's Cathedral, the Barbican and the Guildhall. Two weeks from the middle of July. The Barbican Centre (071 638 4141).

Commonwealth Arts Festival Music and dancing with an international flavour are held in High Street Kensington (071 937 8252).

Notting Hill Gate Carnival On Bank Holiday in August, the Notting Hill Carnival is held in the streets with colourful costumes.

Southwark Shakespeare Festival Performances of Shakespeare plays are held in places around Southwark Cathedral. Tickets and programmes from Festival Office (071 703 2917).

Radio and TV Shows

Free tickets are available to the general public and are obtainable by sending a stamped addressed envelope to the ticket unit of the broadcasting authority listed below stating preference of programme.

British Broadcasting Corporation

A centralised body appointed by the State and responsible to Parliament, though not controlled by the State with regard to policy and programme content. Commenced broadcasting in 1922 with television services following in 1936.

BBC Radio and Television

An extensive range of entertainment and public participation discussion programmes on topical issues are broadcast each week. BBC, Broadcasting House, Portland Place, W1 (071 580 4468).

Independent Broadcasting Authority

Licences are issued to various independent companies in the UK for television and radio. Television services began in 1955 with radio as recently as 1973. Information from **Carlton TV** (071 711 8111), **London Weekend TV** (071 620 1620), **Channel Four Television** (071 631 4444), and **Thames Television** (071 387 4494). LWT building is on the South Bank.

Capital Radio started in 1973. This independent radio station is on the air 24 hours a day broadcasting music mainly from the 1960s and 1970s. Euston Road, NW1 (071 388 1288).

Greater London Radio 35C Marylebone High Street, W1 (071 224 2424).

Essex Radio Radio House, Clifftown Road, Southend-on-Sea (0702 333711).

City of London Churches

The City has fine historic churches, many of which were designed by Sir Christopher Wren. Currently there are over 40 churches which can be visited within the Square Mile. Several of these churches are dedicated to a particular cause or aspect of church work but sadly some are scheduled for closure.

All Hallows, 43 Trinity Square, London Wall, EC3 (071 481 2928). This church was built by George Dance the Younger in 1765-67. The brick west tower has a circular gothic style cupola poised at top. The church was extensively restored 1952-58. Tube Liverpool Street.

St Andrew Undershaft Great St Helens Street, EC3 (071 283 7382).This church was built around its south west tower in 1520-32. The stained glass is of 17th and 19th century make, showing many famous heraldic figures, mainly kings and queens. Tube Aldgate.

St. Botolph Aldgate, EC3 (071 283 1670). Located near the junction of Houndsditch, it was founded before 1291 and rebuilt 1788-91 by Nathaniel Wright. It is in red brick with arched windows and a small west tower. Tube Aldgate.

St Clement Clements Lane, EC4 (071 283 8164). By Wren 1683-7, the church is plain and stuccoed except for the south west tower with exposed brickwork. Tube Monument.

St Edmund the King Lombard Street, EC3 (071 626 9701). By Wren 1670-90, the north tower dates from 1708. The exterior has an italian façade. Tube Bank.

St. Katharine Cree 86 Leadenhall Street, EC3 (071 283 5733). Built 1628-31 in the style of Wren, it was later destroyed and rebuilt with gothic cupola in 1776. Tube Aldgate.

St. Margaret Pattens Eastcheap, EC3 (071 623 6630). By Wren 1684-67 with a simple west front. Tube Monument.

St Martin Ludgate. By Wren 1677-84, the church tower has a lead spire and ogee dome, and balcony. Tube St Pauls.

St. Mary Abchurch EC4 (071 626 0306). Dating to the 12th century, it was rebuilt by Wren 1681-6. There is a lead spire and an ogee domed base. Tube Cannon Street.

St Mary At Hill Lovet Lane, EC3. Originally built by Wren 1670-6, the west end is in yellow brick 1787-8 by George Guilt. Tube Monument.

St Mary-le-Bow, Cheapside, EC2 (071 248 5139). The church dates back to circa1090 and is famous for its bells. Rebuilt by Wren 1670-83, the medieval crypt still has part of the Norman wall dating to the 11th century and its superb Barique Steeple is intact. Tube Mansion House.

St Mary Woolnoth Lombard Street, EC3. Designed by Hawksmoor (1716-27), the church has the most original exterior in London. Tube Bank.

St. Michael Cornhill EC3. Rebuilt by Wren 1670-72 the church tower is original and dates to 1421. Tube Bank.

St Paul's Cathedral (071 248 2705) (see page 44).

St. Stephen, Walbrook, EC4. Founded circa1100, Wren's church lies east of .the original built 1672-9; it is a forerunner of St. Paul's cathedral spacious interior.

St. Vedast, Foster Lane, EC2. Built by Wren 1670-73 and reconstructed after war damage it reopened in 1963. Tube St Pauls.

Dutch Church, Austin Friar, EC2. Founded in 1253 only the nave survived the dissolution and was given to the Dutch protestants. Tube Bank.

Changes proposed by the Templeman Commission include reducing the number of parishes from 22 to 4

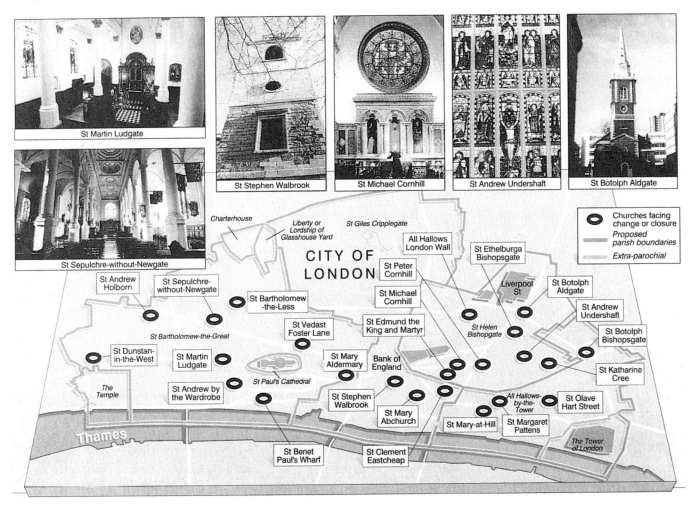

St Martin Ludgate

St Sepulchre-without-Newgate

St Stephen Walbrook

St Michael Cornhill

St Andrew Undershaft

St Botolph Aldgate

Charterhouse

Liberty or Lordship of Glasshouse Yard

St Giles Cripplegate

All Hallows London Wall

St Ethelburga Bishopsgate

Churches facing change or closure

Proposed parish boundaries

Extra-parochial

CITY OF LONDON

St Peter Cornhill

St Michael Cornhill

Liverpool St

St Botolph Aldgate

St Andrew Holborn

St Sepulchre-without-Newgate

St Bartholomew-the-Less

St Edmund the King and Martyr

St Helen Bishopgate

St Andrew Undershaft

St Vedast Foster Lane

St Botolph Bishopsgate

St Bartholomew-the-Great

St Dunstan-in-the-West

St Martin Ludgate

St Mary Aldermary

Bank of England

St Katharine Cree

The Temple

St Andrew by the Wardrobe

St Paul's Cathedral

St Stephen Walbrook

All Hallows-by-the-Tower

St Olave Hart Street

St Mary Abchurch

St Mary-at-Hill

St Margaret Pattens

Thames

St Benet Paul's Wharf

St Clement Eastcheap

The Tower of London

London's Antique, Food and Street Markets

World Antique Centre

The West End of London is a seething commercial centre and clearing house for art scholarship and service. Nowhere else in the world is there so much expertise, so many dealers, collectors, connoisseurs and museums concentrated into such a small area.

Bond Street, Mayfair

Built in the 1880s, this is the fashionable High Street in Mayfair and is packed with art dealers' galleries, and fashion shops.

Christie's and Sotheby's

These have been the world's leading and most elegant auction houses for over two centuries. From their premises a stone's throw apart in Mayfair, Christie's, Sotheby's and Phillips, the world number three, decide the auction trade with the highest Victorian values of Great Britain. Together with the smaller Bonhams the world's number four, they account for three-quarters of world auction sales, their combined total recently reaching £3.2 billion.

General Street Markets

The Arches Villiers Street, WC2

Berwick Street, Wl A well known street market in Soho catering for fruit, vegetables and general articles.

Borough Market, SE17 A fruit and vegetable market in halls under the railway arches of London Bridge Station.

Brixton Market, Atlantic Road, SW19. General market with Caribbean influence.

Caledonia Road Market, Weekend general market, N.1.

Caledonian Market, SE1 An old market in Bermondsey specialising in antiques and open on Fridays only.

Camden Passage, N1 A small market for antiques and shops for fashion clothes in Islington High Street. There is a Boutique Arcade nearby.

Columbia Road (near Aldgate) Flower Market and General goods.

The Cut Waterloo Road, SE1, General market.

Old Covent Garden, WC2 A fashionable and colourful market hall located in the heart of London, and formerly the capital's foremost fruit and vegetable market before that aspect was moved to Nine Elms.

Deptford Market, SE17 General sales market

East Street, SE17 A general market of household items, fruit and vegetables situated in Walworth Road.

Greenwich, SE10 Saturday antiques market.

Hoxton Road Market, N1 Daily market with general goods.

Mile End Waste, E1 General Daily Market opposite London Hospital.

Leather Lane, EC1 Lunchtime Market

Petticoat Lane, E1 An old and colourful market in Middlesex Street containing stalls for clothes, jewellery and many other items of general interest. The Sunday Street market at Petticoat Lane has been a thriving old clothes market since the beginning of the 17th century and it is still a busy general market today for Londoners and visitors. Just a short walk from Liverpool Street station.

Portobello Road, W10 A world famous antique market located in a narrow and crowded street off Westbourne Park.

Ridley Road N8 Fruit and general market.

Roman Road E3 Fruit and general market.

Royal Albert Dock Newly opened Sunday market in London Docklands with a colourful display of general products for sale.

Spitalfields, E1 The historic general market situated in Commercial Street.

Walthamstow, E17 One of the oldest markets in London, it has a mile long narrow High Street packed with stalls and shops.

Specialist Food Markets

Billingsgate, E14 The new Billingsgate Fish Market is situated on the Isle of Dogs.

Leadenhall Market, EC3 A charming Victorian hall, partly for poultry, built in the 1890s and located in the City of London.

New Covent Garden, SW4 The large fruit and vegetable market in Wandsworth, south of the Thames.

Smithfield, EC1 Situated in Charter Street, City of London, this historic and famous wholesale meat centre, is the largest market in the world. Tube Barbican.

Interior of Billingsgate Fish Market

Shops and Shopping Centres Around London

Brent Cross Shopping Centre

London's Department Stores

There were no department stores as we know them in 1850, but there were several large drapers such as Shoolbred's founded in 1820, Swan & Edgar (1812), Dickens & Jones (1803), and Marshall & Snelgrove (1837), and each had more than one department, although it wasn't until the 1860s and 70s that these and other establishments were expanded into larger stores. William Whiteley's (1863), Civil Service Stores (1866), the Army & Navy (1871) and D H Evans (1879) were formed. Harrods which began as a grocery shop developed and became with Whiteley's the two giant Victorian stores. Other stores include Derry & Toms (1920), Barkers (1923), Pontings (1907) and Gamages of Holborn (1878). Woolworth was one of the first chain stores opened in Oxford Street in 1924.

Stores in Central London

There are hundreds of stores in all parts of London where you can buy absolutely anything and a few are listed below from Central London. All these have restaurants and serve good reasonably priced coffees, lunches and afternoon teas.

Barkers Kensington High Street, W8 5SE (071 937 5432) Household, china and many other departments.

British Home Stores Head Office W.1 Marylebone House, 129-137 Marylebone Road NW1 5QD (071 262 3288).
A range of men's, women's and children's clothes. Also household goods, textiles, housewares and many other departments.

W Bill Ltd 28 Old Bond Street W1X 3AB (071 629 9567). Established in 1846, specialise in Cashmere, Shetland and lambswool knitwear for men and women.

C & A 200 Oxford Street, W1N 9DG Offering a range of clothing for the whole family (071 631 4576).

Dickens and Jones Regent Street, W1. The store has American and English styled women's clothes. (071 734 7070).

D.H. Evans Oxford Street, W1.
(071 629 8800) A wide selection of goods including household, linen etc.

Debenhams Oxford Street W1 (071 408 4444). A range of quality merchandise which includes fine china and glassware.

Hamleys Regent Street, W1. A great selection of toys, games and sporting equipment (071 734 3161).

Harrods Knightsbridge, SW1. Quality goods of every description are available and the store is world famous for it's annual sales. (071 730 1234).

Peter Jones Sloane Square, SW1. It has a furnishing fabric department and a boutique. (071 730 3434).

John Lewis Oxford Street, W1. A range of modern goods are available with dress fabric departments (071 629 7711).

Grand Woollen Company Ltd
184/186 Regent Street W1R 5DF (071 437 0970) Cashmere for suits, jackets, etc.

Jaeger 57 Broadwick Street W1V 1FU (071 734 8211). Specialist in classic clothes and knitwear for both men and women.

Marks & Spencer Michael House Baker Street W1A 1DN (071 935 4422). Sells clothes, food, household goods and gifts.

Selfridges Oxford Street, W1. It has stocks of household articles, fashion, furniture, and children's toys (071 629 1234).

The Scotch House 2 Brompton Road Knightsbridge SW1X 7PB (071 581 2151). Clothes made from fine natural fibres for men, women and children.

Docklands Shopping Centres

Hays Galleria SE1 South of the river close to London Bridge Station. The shops include an English herbalist and a Belgian chocolate shop (071 357 7770).

Surrey Quays Shopping Centre SE18 At Docklands largest shopping centre near Surrey Quays tube station, you will find Tesco, BHS, Boots, Ravel, Our Price etc, also a huge free car park.

Tobacco Dock Wapping, E1 In a Grade I listed building along The Highway, this is a speciality retail centre with a Dickensian atmosphere. Car parking facilities.

Shopping Precincts
Brent Cross Shopping Centre
This was the first of the "Hypermarket" type centres and the first undercover centre to be built and opened in 1976. West-End stores have retail premises within this complex.

Hatfield Galleria Straddling the A1 [M] motorway, the Hatfield Galleria has many shops, restaurants, and a multi-screen cinema.

Lakeside Shopping Centre
Lakeside is a giant regional shopping centre situated on the edge of the M25 motorway at junctions 30 and 34, near Dartford Tunnel and Queen Elizabeth II Bridge. The Centre provides free parking for over 30,000 cars.

Shopping Services
Foreign Exchange Company Ltd
169 King Street W6 9JT (071 748 0744). VAT refund service to tourists on behalf of retailers and hotels used.

London Tax Free Shopping Norway House 21-24 Cockspur Street SW1Y 5BN (071 839 4556). No VAT for overseas visitors and tourists.

Sport Stadiums and Associations

Twickenham Rugby Stadium

Wembley Empire Stadium and Arena
Located in Empire Way near Wembley Park and Tube Station, the stadium holds 70,000 people under cover. Famous as the home of the Football Association (FA) Cup Final, it also stages national and international football, hockey, horse riding, the Rugby League final, skating, tennis, basketball and netball in addition to music festivals and concerts. There are also cycle and greyhound races, Horse of the Year Show and Bowling events. (081 902 8833)

Crystal Palace Stadium
Built by the London County Council 1964-66 at Crystal Palace, SE19, it is the first big althetics centre with a stadium seating of 12000. The fine multipurpose sports hall has an 11-storey residential hostel. (081 778 0131)

London Arena
The Arena opened in 1989 on the Isle of Dogs, is the largest sport , entertainment and leisure complex built in London since the establishment of Wembley Stadium in 1924. It has a main area big enough for a soccer match or the world's biggest indoor disco! (071 538 8880)

Ascot Racecourse This was established on Ascot Heath in 1711, and the Royal Ascot Meeting is still one of the major and more glamorous horse race events of the London Racing Season (081 746 1616).

Sports Associations
The sports associations listed here supply information and refer you to a local club. Each of the London Boroughs also provides information and a wealth of leisure facilities. Please consult a local telephone directory.

London Basketball Association
64 Vivian Avenue, Wembley HA9 6RU (081 903 3609).

Amateur Boxing Association Francis House, Francis Street, London SW1P 1DE (071 976 5361).

Cricket Council Lord's Cricket Ground, London NW8 8QN (071 286 4405).

British Cycling Federation 16 Anthony House Nelldale Road, London SE16 2DJ.

British Sports Association for the Disabled 34 Osnaburgh Street, London NW1 3ND (071 383 7229).

Amateur Fencing Association 83 Perham Road, West Kensington, London W14 9SP (071 385 7442).

Football Association 4 Aldworth Grove, Lewisham, London SE13 6HY (081 690 9626).

British Horse Society British Equestrian Centre, Stoneleigh, Kenilworth CV8 2lR (0203 696697).

Keep Fit Association 16 Upper Woburn Place, London WC1H 0QG (071 387 4349).

Lawn Tennis Association Barons Court, W14 (071 385 2366).

All England Netball Association Francis House, Francis Street, London SW1P 1DE (071 828 2176).

Ramblers' Association 1/5 Wandsworth Road, London SW8 2XX (071 582 6878).

Amateur Rowing Association 6 Lower Mall, The Priory, Hammersmith, London W6 9DJ (081 748 3632).

Rugby Football Union Whitton Road, Twickenham TW1 1DZ (081 892 8161).

Clay Pigeon Shooting Association 107 Epping New Road, Buckhurst Hill IG9 5Q (081 505 6221).

National Skating Association of Great Britain Gee Street, London EC1V 3RD (071 253 3824).

Squash Rackets Association West Point, 33 Warple Way, Acton, London W3 0RQ (081 746 1616).

British Water Ski Federation 390 City Road, London EC1V 2QA (071 833 2855).

British Amateur Weight Lifters Association 3 Iffley Turn, Oxford OX4 4DN (0865 778319).

British Amateur Wrestling Association, 60 Calibra Road, N5 (071 226 3931).

Sports Council 16 Upper Woburn Place, London WC1H 0QP (071 388 1277).

International London Marathon Race

The Marathon Race

The marathon is the biggest international event held annually in Britain and is recognised as a championship event in the International Amateur Athletic Federation Calendar. Nearly 30,000 participants are expected to cross the startline at Greenwich in South East London on the third Sunday in April. Over 400 top athletes competing in the World Marathon Cup are drawn from 70 countries. The first-placed man and woman each stand to win more than £25,000.

The Marathon's Route

A recent marathon route is shown covering 26 miles 330 yards. The runners enter the Surrey Quays after some nine miles from Greenwich, cross Tower Bridge at 12 miles and reach the Isle of Dogs via Cable Street, the Commercial Road and West India Dock Road at around 15 miles. After passing eastwards along Aspen Road outside Billingsgate Fish Market, the athletes go through Canary Wharf. At West Ferry Circus, the runners turn left along Marsh Wall and go around the Isle of Dogs. Then the marathon runs down Limehouse, Garnet Street and Wapping High Street and passes along the length of the Highway and in front of Tobacco Dock shopping precinct and St Katharine's Docks. It continues along Upper Thames Street in the City and Victoria Embankment, then passes through Trafalgar Square, onto The Mall, loops in front of Buckingham Palace into Birdcage Walk and finally finishes at Westminster Bridge adjacent to the magestic Houses of Parliament and Big Ben.

Start of the Marathon at Greenwich

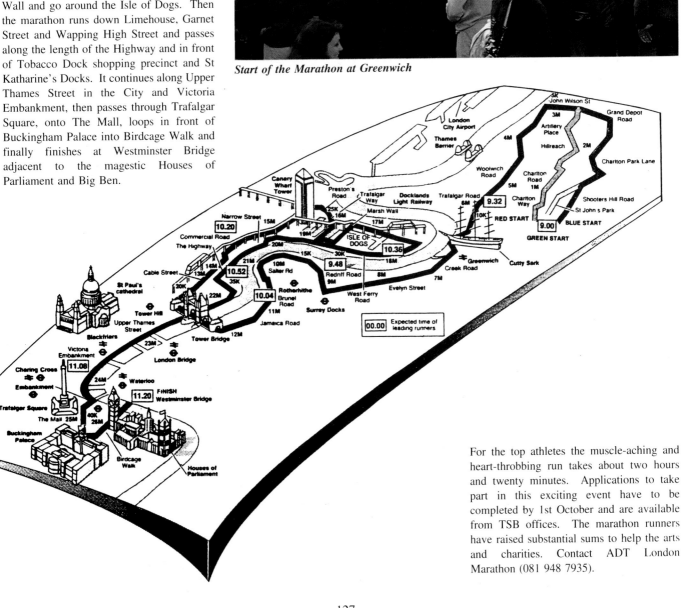

For the top athletes the muscle-aching and heart-throbbing run takes about two hours and twenty minutes. Applications to take part in this exciting event have to be completed by 1st October and are available from TSB offices. The marathon runners have raised substantial sums to help the arts and charities. Contact ADT London Marathon (081 948 7935).

London Local Musuems, Trips & Historic Castles

London Local Museums

Bexley Museum Hall Place, Bourne Road, Bexley, Kent DA5 2PQ (0322 526574 ext 221) BR Bexley.

Bromley Museum The Priory, Church Hill, Orpington BR6 OHH (0689 31551) BR Orpington.

Bruce Castle Museum Lordship Lane, London N17 8NU (081 808 8772).

Burgh House Museum Burgh House, New End Square, London NW3 1LT (081 431 0144) BR Hampstead Heath.

Church Farm House Museum Greyhound Hill, Hendon, London NW4 4JR (081 203 0130) BR West Hendon.

Epping Forest Museum Rangers Road, Chingford, London E4 7QH (081 529 6681) BR Chingford.

Forty Hall Museum Forty Hill, Enfield, Middlesex (081 363 8196) BR Enfield Chase.

Grange Museum of Community History Neasden Lane, London NW10 (081 908 7432) Tube Neasden.

Gunnersbury Park and Museum Gunnersbury Park, London W3 8LQ (081 992 1612) Tube Acton Town.

Harrow Museum and Heritage Centre Headstone Lane, Pinner View, Harrow, Middlesex HA2 6PX (081 861 2626) Tube Harrow on the Hill BR Headstone Lane.

Hogarth House Hogarth Lane, Great West Road, London W4 2QN (081 994 6757) Tube Turnham Green.

Iveagh Bequest, Kenwood Hampstead Lane, London NW3 7JR (081 348 1286) Tube Archway, Golders Green.

Keats House Keats Grove, Hampstead, London NW3 2RR (081 435 2062). Tube Hampstead BR Hampstead Heath

Orleans House Museum Riverside, Twickenham, Middlesex TW1 3DJ (081 892 0221) Tube Richmond.

Pitshanger Manor Museum Mattock Lane, Ealing, London W5 5EQ (081 567 1227) Tube Ealing Broadway.

Valence House Museum Becontree Avenue, Dagenham, Essex RM8 3HT (081 592 4500 ext 4293) Tube Becontree.

Whitehall Cheam 1 Malden Road, Sutton, Surrey SM3 8QD (081 643 1236) BR Cheam

Trips from London

Cheap day trips from London are available to most places, ranging from Brighton on the south coast to the old Roman city of Chester in the north. Trains and coaches leave regularly from British Rail stations and Victoria Coach Station. Most of the trips listed below take about one to two hours each way.

Brighton A lively seaside resort with a continental atmosphere, contains five miles of beaches and two Victorian piers with amusements. BR Victoria.

Cambridge This great university city with its medieval buildings and riverside meadows by the River Cam. BR Liverpool Street.

Canterbury The fine old walled city in Kent is situated on the River Stour and has a magnificent Gothic cathedral, containing the shrine of Thomas a Becket and the tomb of the Black Prince. BR Liverpool Street.

Chatham and Historic Dockyard The English seaport and Royal Naval Base dates from 1588 and was closed in 1984. BR Waterloo.

Colchester This ancient city was England's first Roman town, and has many period relics and remains still on view today. BR Liverpool Street.

Oxford Another university city of beautiful college buildings and church spires, built in local stone and dating back to the 13th century. BR Paddington.

Portsmouth and Naval Museum The naval base houses the Royal Naval Museum. Here also are Henry VIII's Mary Rose, Nelson's H.M.S. Victory and H.M.S. Warrior. BR Waterloo.

Stonehenge, Salisbury and Bath The stone circles at Stonehenge are Britain's most spectacular prehistoric site. By coach or car.

Stratford-on-Avon This Elizabethan town is the birthplace of the great English writer, William Shakespeare (1564-1616). BR Paddington.

Historic Castles and Houses

Arundel Castle (West Sussex) Built at the end of the 11th Century, the castle overlooks the River Arun (0903 883136).

Beaulieu (Hants) The Palace House has been Lord Montague's ancestral home since 1538. It contains the National Motor Museum (0993 811325).

Blenheim Palace (Woodstock, Oxford) It is the birth place of Sir Winston Churchill.

Dover Castle Medieval castle, heavily fortified against the Napoleonic threat.

Hatfield House (Herts) Built 1607-1611, the Jacobean House stands in its park. The staterooms are rich in famous paintings and fine furniture. Elizabethan Banquets are arranged. (0707 262823).

Hever Castle (Hever, Kent) The 13th century moated castle contains fine furniture and works of art (0732 865224).

Leeds Castle (Maidstone, Kent) A royal palace for over three centuries, it contains a superb collection of medieval furniture and tapestries (0622 65400).

HMS Warrior, Portsmouth

City Farms, Zoos, Theme and Nature Parks

City Farms

There are many city farms dotted in and around the capital, ranging from the smallest at Vauxhall to the massive 32 acre Mudchute Farm on the Isle of Dogs. Most are self-supporting and sell their own produce as well as providing a home to a variety of animals. At some of the farms children are invited to bring along their boots and help muck out.

Brook's Farm, Skelton's Lane, E10 (081 539 4278). BR Leyton Midland Road.

College Farm, 45 Fitzalan Road, N3 (081 349 0690). Tube Finchley Central.

Deen City Farm, 1 Batsworth Road, Mitcham (081 648 1461). BR Mitcham.

Freightliners Farm, Sheringham Road, N7 (071 609 0467). Tube Holloway Road/Highbury Islington.

Hackney City Farm, 1a Goldsmiths Row, E2 (071 729 6381). Tube Old Street and 55 bus or Bethnal Green.

Hounslow Urban Farm, Faggs Road, Feltham (081 751 0850). BR Feltham.

Kentish Town City Farm, Cressfield Close, NW5 (071 916 5421). Tube Kentish Town Road or Chalk Farm.

Mudchute Park and Farm, Pier Street, Isle of Dogs, E14 (071 515 5901). Mudchute Station DLR.

Newham City Farm, King George Avenue, E16 (071 476 1170). BR Custom House Victoria Dock (North London Line).

Spitalfields Farm, Weaver Street, E1 (071 247 8762). Tube Whitechapel or Shoreditch.

Stepping Stones Farm, Stepney Way, E1 (071 790 8204). Tube Stepney Green.

Surrey Docks Farm, South Wharf, Rotherhithe Street, SE16 (071 231 1010). Tube Rotherhithe.

Thameside City Farm, 40 Thames Road, Barking (081 594 8449). BR Barking.

Vauxhall City Farm, Tyers Street, SE11 (071 582 4204). Tube Vauxhall.

Wellgate Community Farm, Collier Row Road, Romford (0708 747850). BR Chadwell Heath and bus.

Zoos and Theme Parks

London Zoo in Regents Park

There are more than 12,000 animals to see. It was first opened in 1828 and is now one of the world's foremost animal conservation centres (071 722 3333).

Battersea Park and Children's Zoo

The Park contains a children's zoo, running track and tennis courts. The Veteran Car Run to Brighton starts here (081 871 7530).

Chessington World of Adventure

A 65 acre zoo and circus at Chessington, near Epsom, Surrey, has lions, giraffes, elephants and all the popular wildlife that visitors love to see (0372 727227).

Whipsnade Zoo 500 acres of woodland and parkland located 35 miles north of London with over 2000 animals in large open air enclosures (0582 872171).

Woburn Abbey Zoo A park of 3000 acres located at Woburn in Bedfordshire. Here you can see rare European bison, wallabies, and llamas (0525 290666).

Ada Cole Memorial Stables Located at Broadley Common, near Nazeing in Essex, this sanctuary is a centre for the rescue and care of neglected horses, ponies, donkeys and mules. It is open to the public from 2pm to 5pm every day and admission and parking are free (0992 892133).

Nature Parks

London Wildlife Trust Established in 1981 and affiliated to the Royal Society for Nature Conservation, the Trust manages over forty nature reserves and welcomes school parties for nature studies (071 278 6612).

Camley Street Nature Park The urban park at Kings Cross, has rare blue butterflies, ducks and rich birdlife.

Lavender Pond Nature Park A small man-made pond surrounded by a marsh and woodland in the Surrey Docks.

Stave Hill Ecological Park There is a wildlife trail from Lavender Pond South to Russia Woodland and Stave Hill Park at the centre of the Surrey Docks pennisular.

The Chase This 120 acre park is the largest nature reserve in East London.

Crane Park Island This is a small island in the River Crane which flows through Crane Park, Twickenham.

Havering Country Park

Located in Romford, the park is ideal for birdwatching and the study of flora and fauna. There are miles of horseriding trails with links to Hainault Forest (0708 766999).

Surrey Docks Farm

London Public Houses

Public Houses in London

There are many hundreds of watering holes in London, some dating back to the 16th and 17th century, such as the Prospect of Whitby, London's oldest and most famous riverside Inn at Wapping. More recently new pubs have been established including the Dicken's Inn in St Katharine Docks which was once an old brewery, and is now a free house with two fine restaurants. There is also the Town of Ramsgate where you might discover a piece of history - there is a pirate gallows in the yard! A few pubs in Central London are listed below.

The City of London

Barley Mow, 50 Long Lane, EC1 (071 606 6591) 16th Century free house.

Blackfriar, 174 Queen Victoria Street, EC4 (071 236 5650).

Bull's Head, 80 Leadenhall Street, EC3 (071 283 2830).

Cheshire Cheese, Crutched Friars, EC3 (071 481 1533) Famous old pub.

Fox and Anchor, 115 Charterhouse Street, EC1 (071 253 4838). Built in 1789.

Jamaica Wine House, St. Michael's Alley, Cornhill, EC3 (071 626 9496).

Magogs, 8 Russia Row, EC2 (071 606 3293). Modern round pub.

Pavilion End Pub, 23 Watling Street, EC4 (071 236 6719. Cricket themed pub.

Ye Olde Mitre Tavern, 1 Ely Place, EC1 (071 405 4751). Dates back to 1546.

Bloomsbury and Holborn

Cittie of York, 22 High Holborn, WC1 (071 242 7670. Large 17th Century pub.

Museum Tavern, 49 Gt. Russell Street, WC1 (071 242 8987).

Ship Tavern, 12 Gate Street, WC2 (071 405 1992). Dates back to 16th Century.

White Lion, 15 St Giles High Street (071 836 8956). At rear of Centre Point Tower.

Bayswater and Paddington

Archery Tavern, 4 Bathhurst Street, W2 (071 402 4916). Archery pictures.

Mitre, 24 Craven Terrace, W2 (071 262 5240). Has cellar wine bar.

Sussex, 21 London Street, W2 (071 402 9602). Convenient for Paddington Station.

Camden

Grand Junction Arms, Acton Lane NW10 (071 965 5670).

Waterside Inn, 82 York Way, Kings Cross, N1 (071 837 7118).

Covent Garden

Freemason's Arms, 81 Long Acre, WC2 (071 836 6931). Beamed ceiling.

Lamb and Flag, 33 Rose Street, WC2 (071 836 4108). 300 year old pub.

Lemon Tree, 4 Bedfordbury, WC2 (071 386 1864). Convenient for the market.

Marquis of Anglesea, 39 Bow Street,

Courage Brewery Shire Horses

WC2 (071 836 3216). 18th century origin.

Opera Tavern, 23 Catherine Street, WC2 (071 836 7321). Opposite The Royal Theatre.

White Hart, 1 New Row, WC2 (071 836 3291). Old pub in Covent Garden.

Fleet Street and Strand

Cartoonist, 76 Shoe Lane, EC4 (071 353 2828). Headquarters of Cartoonist Club.

Cheshire Cheese, 145 Fleet Street, EC4 (071 353 6170). 14th century origins.

Cock Tavern, 22 Fleet Street, EC4 (071 353 8570). Dickensian mementoes.

Edgar Wallace, 40 Sussex Street, WC2 (071 353 3120).

The George, Fleet Street, EC4 (071 353 9238). Opposite the Law Courts.

Old Bell Tavern, 95 Fleet Street, EC4 (071 353 3796). Built 1670 by Christopher Wren.

Knightsbridge & Kensington

Bunch of Grapes, 207 Brompton Road, SW3 (071 589 4944). Victorian interior.

Churchill Arms, 119 Kensington Church Street, W8 (071 727 4242).

Duke of Cumberland, 235 New King's Road, SW6 (071 736 2777).

Star Tavern, 6 Belgrave Mews West, SW1 (071 235 3019). Open fires in the bars.

Leicester Square and Soho

Blue Posts, 28 Rupert Street, W1 (071 437 1415). Oil paintings.

Coach and Horses, 1 Great Marlborough Street, W1 (071 437 3282). 18th century Inn.

Cumberland Stores, 15 Beak Street, W1 (071 734 5870). American food.

Dog and Duck, 18 Bateman Street, W1 (071 437 3478). 18th century pub.

Duke of Wellington, 77 Wardour Street, W1 (071 437 2886). Old pub.

The French House, 49 Dean Street, W! (071 437 2799). Famous people drank here.

Mayfair

Burlington Bertie, 21 Old Burlington Street, W1 (071 437 8355).

Red Lion, Waverton Street, W1 (071 499 1307). 17th century inn.

Rose and Crown, 2 Old Park Lane, W1 (071 499 1980). Said to be haunted!

Westminster and Whitehall

Buckingham Arms, 62 Petty France, SW1 (071 222 3386).

Clarence, 53 Whitehall, SW1 (071 930 4808). 18th century house.

St. Stephen's Tavern, 10 Bridge Street, SW1 (071 930 3230). MPs' local.

Silver Cross, 33 Whitehall, SW1 (071 930 8350). 13th century origin.

Westminster Arms, 9 Storey's Gate, SW1 (071 222 8520). Queen Anne Bar.

East London Pubs

Blind Beggar Public House

East London is full of old pubs with old tales of customers like Dickens, Pepys and Nelson. The Blind Beggar Pub in Whitechapel has been the scene of both murder and salvation. This is the site where Ronnie Kray shot his arch-enemy George Cornell in 1966 and recently has been the location for the movie "The Krays". In contrast, the pub marks the birth of the Salvation Army. During the 1880s General Booth stood outside the pub and sold the first ever copy of 'War Cry' the Salvation Army newspaper.

London Theatres, Night Clubs and Concert Halls

West End Night Life

The West End in Central London offers a dazzling choice of theatres, night clubs, cinemas, wine bars, cafes and a considerable number of restaurants offering English and international cuisine. For nightspots, visit the Cabarets and Music Halls around Piccadilly Circus. Besides a good meal there is spectacular music hall variety, hilarious magic, can-can and dancing until late. Such establishments include The Rock Circus, the Cockney and Talk of the Town.

Theatres

Aldwych Home of the Royal National Theatre (071 836 6406).

Apollo Shaftesbury Avenue, W1, musicals and farces (071 494 5070).

Comedy In Panton Street, SW1, comedy plays and musicals (071 867 1045).

Coliseum St Martins Lane, WC2 (071 836 3161.

Drury Lane Catherine Street, WC2 (071 494 5001).

Duke of York Shows plays and comedies (071 836 5122).

Garrick Charing Cross Road, WC2, plays and musicals (071 494 5085).

Haymarket SW1, classic productions and plays (071 930 8800).

Her Majesty's SW1, musical plays (071 494 5400).

London Palladium Argyll Street, W1 (071 494 5020).

Lyric Shaftesbury Avenue, W1 specialises in musical productions (071 494 5045).

Mermaid Upper Thames Street, EC4 (071 410 0000).

Old Vic Waterloo Road, SE1, houses the National Theatre (071 928 2651).

Prince Edward Compton Street, W1 Musical plays (071 734 8951).

Prince of Wales Theatre Coventry Street, W1, musicals and plays (071 839 5972).

Strand Theatre Aldwych, WC2 (071 836 4143).

Victoria Palace Victoria Street, SW1 revues etc.(071 834 1317).

Wyndham's Charing Cross Road, WC2, thrillers, plays and musicals (071 867 1125).

Concert Halls

Barbican Hall It is the home of the London Symphony Orchestra and Royal Shakespeare Company (071 638 4141).

Caxton Hall Built in 1883 as the Town Hall of the City of Westminster, today it is used for concerts, weddings and public meetings (071 222 5212).

Conway Hall The Headquarters of the Methodist Church in Storeys Gate, SW1, organ recitals, orchestral concerts and public meetings are held here (071 242 8032).

Royal Opera House It is the home of great ballet and opera located in Bow Street, Covent Garden, WC2 (071 240 1066).

Queen Elizabeth Hall The concert hall of South Bank Centre, SE1 (071 928 8800).

Royal Albert Hall The hall in Kensington Gore, SW7, is famous for the annual 'Proms' and public meetings (071 589 8212).

Royal Festival Hall With a seating of 3000, it is famous for its orchestral and choral concerts (071 928 8800).

Sadlers Wells The home of fine ballet and opera for over a hundred years, is situated in Rosebery Avenue, EC1 (071 278 8910).

Wigmore Hall The hall at 36 Wigmore Street, W1, holds chamber music and song recitals (071 935 2141).

Night Clubs/Discotheques

Broadway Boulevard The High Street, Ealing Broadway, W5 (081 840 0616).

Equinox Discotheque Leicester Square, WC2 (071-437 1446).

The Hippodrome Hippodrome Corner, Leicester Square, WC2 (071 437 4311).

The 100 Club 100 Oxford Street, WC1 (071 636 0933).

The Rock Circus, Shaftesbury Avenue, W1 (071 734 8025).

The Stork Club 99 Regent Street, W1 (071-734 3686).

Evening Out and Cabaret

The Beefeater by the Tower, Ivory House, E1 (071 480 7017).

Talk of London Cabaret, Drury Lane WC2 (071 405 1516).

The Cockney, 161 Tottenham Court Road, W1 (071 263 0343).

West End Cinemas

Cannon	Haymarket	071 839 1527
Cannon	Oxford St.	071 636 0310
Cannon	Piccadilly	071 437 3561
Cannon	Royal Charing X Rd	071 930 6915
Cannon	Premiere Leics Sq.	071 439 4470
Cannon	Shaftesbury Ave.	071 836 6279
Cannon	Tottenham Ct.Rd	071 636 6148
Cannon	Moulin Windmill St.	071 437 1653
Curzon	Phoenix Ch X Rd	071 240 9661
Curzon	West End Shafts Ave.	071 439 4805
Dominion	Tott Ct.Rd.	071 580 9562
Empire	Leicester Sq.	071 437 1234
Lumiere	St. Martin's Lane	071 379 3014
Metro	Rupert St.	071 437 0757
Odeon	Haymarket	071 839 7697
Odeon	Leicester Sq.	071 930 6111
Plaza	Piccadilly Circus	0800 888997
Prince Charles	Leics Sq.	071 437 8181
Screen	Baker St.	071 935 2772
Warner	Leicester Sq.	071 439 0791

Theatre Tickets & Sport

London Ticket Bureau, 60 Gloucester Terrace, W2 (071 937 0005).

London Palladium

London Hotels, Celebrations & Weekend Breaks

New Year Celebrations

The favourite place to celebrate the New Year in London is Trafalgar Square. There is also a lesser gathering in nearby Parliament Square beside Big Ben. On New Year's Day there is a parade with many musicians, dancers, acrobats, clowns and floats. It starts from Parliament Square at noon. Oxford and Regents Streets have beautiful lights until the 6th January. For bargain hunters, Selfridges in Oxford Street starts its sale directly after Boxing Day but Harrods waits until the first week in January. All the big shops open on New Year's day for their sales.

To celebrate the New Year in rich style, you can stay in one of the luxury hotels in the West End such as Claridges in the heart of Mayfair. It is one of London's traditional hotels. Two orchestras and a piper normally play at the New Year's Eve party where you can have a superb champagne banquet (071 629 8860). Drury Lane Moat House in Covent Garden has a New Year's Eve Theatre Break including dinner, disco and a matinee on New Year's Day (071 836 6666). During the Christmas period, Sadler's Wells normally presents Peter Pan (071 278 8916). The Nutcracker, a favourite every year, is staged by the English National Ballet at the Royal Festival Hall at South Bank (071 928 8800). Central Line coaches bring people to London from fourteen places around the West Midlands, early on New Year's Eve to see a musical followed by supper and a disco (021 333 3232). Individually, you can book accommodation with a credit card through the London Tourist Board on (071 824 8844). For more information phone Visitorcall on (0839 123456). A one-day Travelcard is available after 9.30am from underground stations and some newsagents for use on tubes and buses.

West End Hotels

Athenaeum Hotel Piccadilly, W1V 0BJ (071 499 3464).

Bloomsbury Coram Street, WC1N 1HT (071 837 1200).

Cumberland Hotel Marble Arch, London, W1A 4RF (071 262 1234).

Grosvenor Park Lane, W1 3AA (071 499 6363).

Hotel Russell Russell Square, WC1B 5BE (071 837 6470).

Hyde Park Hotel 66 Knightsbridge, SW1Y 7LA (071 235 2000).

Regent Palace Hotel P O Box 4BZ, Piccadilly Circus, W1A 4BZ (071 734 7000).

Regents Park Carburton Street, W1P 8EE (071 388 2300).

Ritz Hotel Piccadilly, W1 (071 493 8181). Popular restaurant for that special occasion.

Royal Lancaster Hotel Lancaster Terrace, W2 2TY (071 262 6737).

St. James's 81 Jermyn Street, SW1Y 6JF.

Strand Palace 372 The Strand, WC2R 0JJ (071 836 8080).

Waldorf Aldwych, WC2B (071 836 2400).

Weekend Breaks

Bonnington Hotels 92 Southampton Row, WC1 (071 242 2828).

Carnarvon Hotel Ealing Common, W5 (081 992 5399).

Crest Welcome Breaks At West End Hotels (0295 67722).

National Express 4 Vicarage Road, Birmingham, B15 (021 456 1102).

Selsdon Park Hotel Sanderstead, South Croydon, Surrey CR2 8YA (081 657 8811). Weekend breaks including golf and tennis.

Trusthouse Forte Hotels 24-30 New Street, Aylesbury, Bucks, HP20 2NW (081 567 3444). Twenty hotels are available.

Sightseeing Tours

Cityrama Silverthorne Road, Battersea, SW8 (071 720 6663). Regular daily coach tours around London.

Evan Evans Tours 26 Cockspur Street, Trafalgar Square, SW1 (Booking 081 332 2222). Extensive sightseeing coach tours around London and Britain.

Golden Tours 11 Poplar Mews, Uxbridge Road, W12 (071 743 3300). Comprehensive range of coach tours using Blue Badge guides.

Rock Tours (071 734 0227) London's rock music tours commencing in Regent Street.

Visitors Sightseeing 35/36 Woburn Place, WC1 (071 636 7175).

Frames Richards 11 Herbrand Street, WC1 (071 837 3111).

National Express Victoria Station, SW1 (071 730 0202).

London Coaches Jews Row, SW18 (071 828 7395).

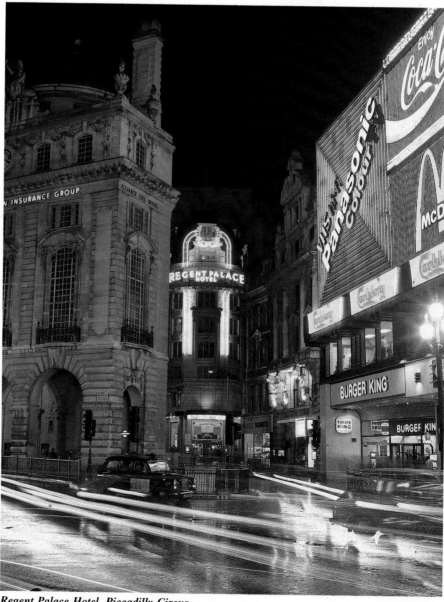

Regent Palace Hotel, Piccadilly Circus

Travel and Sightseeing Information

Rolls Royce - travel in style!

London Airports

Heathrow Airport,Bath Road, Heathrow, Middlesex (081 759 4321).
London City Airport King George V Dock E.16 (071 474 5555).
Gatwick Airport Crawley (0293 535353).
Stansted Airport, Essex (0279 680500).
Luton Airport, Luton (0582 395000).

Air Terminal

Victoria Station Air Terminal
Victoria Station, SW1 (071 834 9411).

Forwarding Agents

London Baggage Co, 262 Vauxhall Bridge Road SW1V 1BB (071 828 2400).
Personal Shipping Services
35 Craven Street WC2 (071 407 6606).

Car Rental

Avis Rent-a-Car Trident House, Station Road, Hayes, Middlesex (081 848 8733).
EuroCar Bushey, Herts (0932 819000).
Hertz Rent-A-Car 35 Edgware Road, Marble Arch, W2 2JE (071 402 4242).
EuroDollar
305 Chiswick High Road, W4 4HH (0895 233300).

Ferries to Europe

Brittany Ferries (0705 827701).
Hoverspeed Ltd (0304 240101).
P & O European Ferries (0304 203388).

Note to Reader Telephone numbers and names in this book could change. Please contact telephone directory or consult London Tourist Board.

Sightseeing Information

London Tourist Board
Victoria, Station Parade SW1 (071 730 3488).
British Travel Service
54 Eubury Street, SW1 (071 730 8986).
City of London Information Centre
St Paul's Churchyard, EC4 (071 606 3030).
Docklands Visitor Centre 3 Limeharbour, E14 (071 512 3000).
National Trust For places of historic interest or natural beauty (081 464 1111).
English Heritage For historic places, castles etc. (071 973 3000).

Taxis

Licensed Taxi Drivers Association 9/11 Woodfield Road, W9 (071 286 1046).

Coaches

**Western Coaches and
A1 Luxury Coaches of London**
98 High Road, Woodford Green, Essex IG8 9EF (081 504 9747).
Evan Evans Tours Limited
27 Cockspur Street, Trafalgar Square, SW1Y 5BT (071 839 6415).
Green Line Coaches Lesbourne Road, Reigate, Surrey RH2 7LE (0737 242411)

London Transport

Full information and 24 hour telephone service: 55 The Broadway, Victoria, SW1 (071 222 1234)

British Rail

Euston Station (071 387 7070), Kings Cross (071 278 2477), Paddington (071 262 6767), Waterloo and Victoria (071 928 5100).

AA Roadwatch

Traffic, Roadworks and Weather
Phone 0839 500 591
National Motorways
Phone 0836 401100

Index

Page references for illustrations are in italic wherever the reference differs from that of the subject matter. Abbreviations: G=Gallery, Ga=Garden, M=Museum, P=Park.

British Museum and London University

Heathrow Airport with shopping complex

Acknowledgements

The Royal Albert Hall

The Albert Memorial

I wish to record my gratitude to the University of East London for the support of the research work over many years. I gratefully acknowledge the debt I owe to many individuals, previous writers, photographers and diverse organisations who so kindly helped with the preparation of this publication. For the generous supply of excellent aerial photographs, including the cover illustrations and the permission to reproduce, I am deeply grateful to Chorley and Handford. The assistance and co-operation of Tom Samson and Paul Proctor in this respect have been invaluable. Special thanks are due to Sir Francis McWilliams, a former Mayor of London, and the staff of the Corporation of London for the supply of information. Acknowledgement has to be made of the considerable help received from Jeremy Smith of the Guildhall Library and Helen McCrorie of the Public Relations Department, Jenny Hall and Gavin Morgan of the Museum of London have kindly supplied maps, reports and slides.

Much appreciation is due to the following who checked the manuscript and made many helpful revisions: John Noble, Ron McDougall, Paul and Janice Smith, Ian Smith, Loreta Fleming, Ted Weedon, Ian Williams, and Roy Lepley. The index was expertly prepared by John Noble. For assistance with the historical research and external contacts I am deeply grateful to Terence O'Connell. Thanks are due to Simon Pattie who greatly helped with the preparation of slides and to Gary Jewell for his general assistance.

For the supply of information and in some cases transparencies for the illustrations, I am deeply grateful to many organisations. They include The Royal Collections by gracious permission of Her Majesty Queen Elizabeth II, Royal Botanic Gardens at Kew, Forte Hotels, Syon House, Westminster Abbey, English Heritage, National Gallery, National Portrait Gallery, Department of the Environment, London Underground, British Rail Network Southeast, New Civil Engineer, Port of London Authority and their Magazine, St Paul's Cathedral, Southwark Cathedral, All Hollows by the Tower, Science Museum, London Docklands Development Corporation, St Martin Property Company, London Borough of Richmond Upon Thames, Portsmouth Borough Council, Selfridges' Archive, The Daily Telegraph, The Times Newspapers, Olympia & York, British Museum, HM Tower of London, London Tourist Board, English Tourist Service, Evan Evans Tours, St Katherine by the Tower, Imperial War Museum, former Rank Hotels, William Morris Gallery, Savills, Prudential Property Services, BICC Group, National Maritime Museum, National River Authority, British Waterways, Thames Water, the Livery Companies, Hornsey Historical Society, London Canal Museum, and Trafalgar House.

I would like to express my thanks to Tom Juffs for constant help and valuable advice, to Derek Merritt for his kind support, Michael Paule for tracing of maps, Rudra Vasanthakumaran for help with preparation of maps, Linda Day for excellent typing and patient revision, Alan Hooker for continuous advice on the use of the wordprocessor, Philip Jupps for artwork, Barry Nottage for library research, Sheila Johnson for administration of research grants, Gordon Graham for his kind advice and help with the printing and Librarians of the London Boroughs who sent information. Thanks are also due to Joanna Maddison, Margaret Youngman, Sunny Crouch, Geoff Nyberg, Jane Ellis, Terry Hatton, Roger Mutton, David King, Sue Ellison, Fred Redding, Grant Smith, Lucy Sapte, Alf Lucas, Patricia O'Connell, Isobel Stokes, Carole Anthony, Ruth Thraves and Philip Porter.

Many thanks are due to John Jones for advice, expert artwork and co-operation. I also thank my wife Irene, for her continued help and lasting patience over many years. To the general public and visitors, who have generously supported my books, I express my deepest appreciation.

London attractions